METHODS IN MOLECULAR BIOLOGY™

Series Editor
John M. Walker
School of Life Sciences
University of Hertfordshire
Hatfield, Hertfordshire, AL10 9AB, UK

For other titles published in this series, go to
www.springer.com/series/7651

Microbial Toxins

Methods and Protocols

Edited by

Otto Holst

Forschungszentrum Borstel, Leibniz-Zentrum für Medizin und Biowissenschaften, Parkallee 4a/c, 23845, Borstel, Germany

Editor
Otto Holst
Forschungszentrum Borstel, Leibniz-Zentrum für
Medizin und Biowissenschaften
Parkallee 4a/c
23845, Borstel
Germany
otto.holst@googlemail.com

ISSN 1064-3745 e-ISSN 1940-6029
ISBN 978-1-61779-101-7 e-ISBN 978-1-61779-102-4
DOI 10.1007/978-1-61779-102-4

Springer New York Dordrecht Heidelberg London

Library of Congress Control Number: 2011928553

Printed on acid-free paper

Humana Press is part of Springer Science+Business Media (www.springer.com)

Preface

In the year 2000, a first methods collection entitled *Bacterial Toxins: Methods and Protocols*, which contained 20 chapters on protein toxins and endotoxin from bacteria and cyanobacteria, was published. The idea was to support researchers of various scientific disciplines with detailed descriptions of state-of-the-art protocols and, since the book turned out to be quite successful, it is quite obvious that this aim could be achieved. However, it was also noticed that by focusing on bacterial toxins one significant area of toxin research was missing, namely, mold fungus toxins. Now, 10 years later, a second volume entitled *Microbial Toxins: Methods and Protocols*, which includes protocols on mold fungus toxins with some focus on aflatoxins, is presented.

The interest of researchers across a broad spectrum of scientific disciplines in the field of microbial toxins is clearly unbroken. Since this field, as many others do, makes use of a large variety of biological, chemical, physical, and medical approaches, investigators dealing with any microbial toxin have to be familiar with a number of techniques from all these disciplines. The book *Microbial Toxins: Methods and Protocols* intends to strongly support researchers here.

Microbial Toxins: Methods and Protocols consists of 20 chapters classified into three sections *I*, Bacterial Protein Toxins, *II*, Endotoxins, and *III*, Mold Fungus Toxins. The protocols collected represent state-of-the-art techniques which are described by authors who have regularly been using the protocol in their own laboratories. In each chapter, a brief introduction to the method being described is included, followed by a step-by-step description of the method treated. Each chapter also possesses a Notes section in which, e.g., difficulties, modifications, and limitations of the techniques are exemplified. In sum, our volume, *Microbial Toxins: Methods and Protocols*, should prove useful to many researchers, including those without any previous experience with a particular technique.

Borstel, Germany ***Otto Holst***

Contents

Contributors

AZIZ AMINE • *Dipartimento di Scienze e Tecnologie Chimiche, Università di Roma Tor Vergata, Roma, Italy*
FABIANA ARDINI • *Dipartimento di Scienze e Tecnologie Chimiche, Università di Roma Tor Vergata, Roma, Italy*
KARINE BAGRAMYAN • *Department of Immunology, The Beckman Research Institute of the City of Hope, Duarte, CA, USA*
FERNANDO BATTAGLINI • *INQUIMAE – DQIAQF, Universidad de Buenos Aires, C1428EHA, Buenos Aires, Argentina*
MARTINE CAROFF • *Equipe "Structure et Activités des Endotoxines", Institut de Génétique et Microbiologie, Université de Paris Sud-XI, Orsay, France*
DIEGO CENTONZE • *Dipartimento di Scienze Agro-Ambientali, Chimica e Difesa Vegetale and BIOAGROMED, Università degli Studi di Foggia, Foggia, Italy*
PRABIR K. DUTTA • *The Department of Chemistry, Ohio State University, Columbus, OH, USA*
TERESA GONZÁLEZ-JAÉN • *Department of Genetics, Universidad Complutense de Madrid, Madrid, Spain*
MARIA GONZALEZ-MOA • *Center for Innovations in Medicine, The Biodesign Institute at ASU, Tempe, AZ, USA*
AMAIA GONZÁLEZ-SALGADO • *Department of Genetics, Universidad Complutense de Madrid, Madrid, Spain*
MARCY HERNICK • *Department of Biochemistry, Virginia Tech, Blacksburg, VA, USA*
MARCO IAMMARINO • *Istituto Zooprofilattico Sperimentale della Puglia e della Basilicata, Foggia, Italy*
MARKUS KALKUM • *Department of Immunology, The Beckman Research Institute of the City of Hope, Duarte, CA, USA*
ANIKÓ KILÁR • *Faculty of Medicine, Institute of Bioanalysis, University of Pécs, Pécs, Hungary*
FERENC KILÁR • *Faculty of Medicine, Institute of Bioanalysis, University of Pécs, Pécs, Hungary*
BÉLA KOCSIS • *Faculty of Medicine, Institute of Medical Microbiology and Immunology, University of Pécs, Pécs, Hungary*
KRISZTINA KOVÁCS • *Faculty of Medicine, Institute of Medical Microbiology and Immunology, University of Pécs, Pécs, Hungary*
SONIA LO MAGRO • *Istituto Zooprofilattico Sperimentale della Puglia e della Basilicata, Foggia, Italy*
LILLA MAKSZIN • *Faculty of Medicine, Institute of Bioanalysis, University of Pécs, Pécs, Hungary*

SHANE MASSEY • *Biomolecular Research Annex, Orlando, FL, USA*
CARLOS MORALES-BETANZOS • *Center for Innovations in Medicine, The Biodesign Institute at ASU, Tempe, AZ, USA*
DANILA MOSCONE • *Dipartimento di Scienze e Tecnologie Chimiche, Università di Roma Tor Vergata, Roma, Italy*
MARILENA MUSCARELLA • *Istituto Zooprofilattico Sperimentale della Puglia e della Basilicata, Foggia, Italy*
DONATELLA NARDIELLO • *Istituto Zooprofilattico Sperimentale della Puglia e della Basilicata, Foggia, Italy*
ALEXEY NOVIKOV • *Equipe "Structure et Activités des Endotoxines", Institut de Génétique et Microbiologie, Université de Paris Sud-XI, Orsay, France*
CARMEN PALERMO • *Dipartimento di Scienze Agro-Ambientali, Chimica e Difesa Vegetale and BIOAGROMED, Università degli Studi di Foggia, Foggia, Italy*
DIEGO PALLAROLA • *INQUIMAE – DQIAQF, Universidad de Buenos Aires, C1428EHA, Buenos Aires, Argentina*
BELÉN PATIÑO • *Department of Microbiology III, Universidad Complutense de Madrid, Madrid, Spain*
ANDREW J. PHIPPS • *Department of Veterinary Biosciences, The Center for Microbial Interface Biology, Ohio State University, Columbus, OH, USA*
GABRIELLA POCSFALVI • *Institute of Protein Biochemistry, National Research Council, Naples, Italy*
ELDER PUPO • *Department of Vaccinology, National Institute for Public Health and the Environment (RIVM), Bilthoven, The Netherlands*
BEATRIZ QUIÑONES • *United States Department of Agriculture/Agricultural Research Service, Produce Safety and Microbiology Research Unit, Western Regional Research Center, Albany, CA, USA*
ANNE RANTALA-YLINEN • *Division of Microbiology, Department of Food and Environmental Sciences, University of Helsinki, Helsinki, Finland*
RONALD T. RILEY • *USDA – ARS, Toxicology and Mycotoxin Research Unit, R.B. Russell Research Center, Athens, GA, USA*
GITTA SCHLOSSER • *Research Group of Peptide Chemistry, Hungarian Academy of Sciences, Eötvös L. University, Budapest, Hungary*
WILLIAM C. SCHUMACHER • *The Department of Chemistry, Ohio State University, Columbus, OH, USA*
HANNA SIPARI • *Division of Microbiology, Department of Food and Environmental Sciences, University of Helsinki, Helsinki, Finland*
KAARINA SIVONEN • *Division of Microbiology, Department of Food and Environmental Sciences, University of Helsinki, Helsinki, Finland*
CRAIG A. STOROZUK • *Department of Veterinary Biosciences, The Center for Microbial Interface Biology, Ohio State University, Columbus, OH, USA*
SERGEI A. SVAROVSKY • *Center for Innovations in Medicine, The Biodesign Institute at ASU, Tempe, AZ, USA*
MICHELLE S. SWIMLEY • *United States Department of Agriculture/Agricultural Research Service, Produce Safety and Microbiology Research Unit, Western Regional Research Center, Albany, CA, USA*

KEN TETER • *Biomolecular Research Annex, Orlando, FL, USA*
COVADONGA VASQUEZ • *Department of Microbiology III, Universidad Complutense de Madrid, Madrid, Spain*
WATARU YAMAZAKI • *Faculty of Agriculture, Department of Veterinary Science, University of Miyazaki, Miyazaki, Japan*
NICHOLAS C. ZITOMER • *USDA – ARS, Toxicology and Mycotoxin Research Unit, R.B. Russell Research Center, Athens, GA, USA*

Part I

Bacterial Protein Toxins

Chapter 1

Detection of Bacterial Protein Toxins by Solid Phase Magnetic Immunocapture and Mass Spectrometry

Gabriella Pocsfalvi and Gitta Schlosser

Abstract

Bacterial protein toxins are involved in a number of infectious and foodborne diseases and are considered as potential biological warfare agents as well. Their sensitive multiplex detection in complex environmental, food, and biological samples are an important although challenging task. Solid-phase immunoaffinity capture provides an efficient way to enrich and purify a wide range of proteins from complex mixtures. We have shown that staphylococcal enterotoxins, for example, can be efficiently enriched by means of magnetic immunocapture using antibody functionalized paramagnetic beads. The method was successfully interfaced by the on-beads and off-beads detection using matrix-assisted laser desorption/ionization time-of-flight mass spectrometry at the protein level and by the off-beads nano-electrospray ionization-MS/MS detection at the enzyme digests level, enabling thus the unambiguous identification of the toxin. The method is applicable to any bacterial toxin to which an antibody is available.

Key words: Bacterial protein toxins, Staphylococcal enterotoxins, Immunomagnetic separation, Immunoaffinity, Magnetic beads, Mass spectrometry, MALDI, ESI

1. Introduction

Mass spectrometry (MS) has been successfully applied in the detection, identification, quantification, and structural characterization of biological toxins (1, 2), including bacterial protein toxins, such as botulinum (3–7), shiga-like (8), tetanus (9), anthrax (10) toxins, and staphylococcal enterotoxins (SEs) (2, 11). Both matrix-assisted laser desorption/ionization (MALDI) (12) and electrospray ionization (ESI) (9) techniques have been widely exploited for the characterization of protein toxins. Their direct detection in complex biological samples like body fluids and food

Otto Holst (ed.), *Microbial Toxins: Methods and Protocols*, Methods in Molecular Biology, vol. 739,
DOI 10.1007/978-1-61779-102-4_1, © Springer Science+Business Media, LLC 2011

is hampered by the generally low concentration of protein toxin and by the suppression effect of the matrix. Sensitivity and specificity of these techniques can be enhanced by coupling the MS-based detection (1) to the measurement of specific enzymatic activity of the toxin by the detection/quantification of its product peptides (7), (2) to immunoaffinity enrichment/purification step (11), or by the combined application of the two (6, 10). Activity-based MS was proven to be extremely sensitive for the detection of those bacterial protein toxins which have highly specific protease activity like botulinum and anthrax toxins. Immunoaffinity enrichment, on the other hand, can be used independently from the biological activity in an unbiased way. Today, highly specific antibodies are available for a major part of bacterial protein toxins (13), which make the immunoaffinity enrichment-based MS method feasible. One of the most convenient ways to perform affinity capture is the application of antibody-coated magnetic particles as affinity probe. The major advantage of magnetic beads is that after the enrichment process they can easily be separated from complex matrices and are amenable to downstream MS-based analysis. In this chapter, we describe a procedure for the preparation of SEs immunomagnetic affinity probes and their use in MS-based detection. The method is generally applicable to other bacterial protein toxins to which antibodies are available.

2. Materials

2.1. Preparation and Use of Antibody-Coated Magnetic Particles

1. Lyophylized SEs from Toxin Technology, Inc., Sarasota, FL, USA are dissolved at 1 mg/mL in water and stored in aliquots at −20°C. SEs toxins are particularly toxic (see Note 1).
2. Polyclonal, affinity purified anti-staphylococcal enterotoxin IgGs, ≥95% IgG, from Toxin Technology, Inc. Sarasota, FL, USA, are dissolved at 1 mg/mL in water and stored in aliquots at −80°C.
3. Magnetic particles: Dynabeads® M-280, tosylactivated superparamagnetic polystyrene beads coated with a polyurethane layer are used from Dynal, Norway. Physical characteristics: diameter: 2.8 μm ± 0.2 μm, surface area: 4–8 m^2/g, active chemical functionality: 50–70 μmol/g, density: 1.3 g/cm^3, concentration: 2×10^9 beads/mL (approximately 30 mg/mL). Store at 4°C.
4. Supernatant removal and washing steps are performed using a magnetic particle concentrator from Dynal, Norway.
5. Buffer A: 0.1 M Na-phosphate buffer, pH 7.4 (2.62 g $NaH_2PO_4 \cdot H_2O$ and 14.42 g $Na_2HPO_4 \cdot 2H_2O$ dissolved in water and volume adjusted to 1,000 mL). Store at 4°C.

6. Buffer B: phosphate-buffered saline (PBS), pH 7.4, with 0.1% (w/v) bovine serum albumin (BSA) (0.88 g NaCl and 0.8 g BSA, dissolved in 0.01 M Na-phosphate pH 7.4 and volume adjusted to 100 mL). Store at 4°C.
7. Buffer C: 0.2 M Tris–HCl pH 8.5 with 0.1% (w/v) BSA (2.42 g Tris–HCl dissolved in distilled water, pH adjusted with 1 M HCl to 8.5, and volume adjusted to 100 mL). Store at 4°C.
8. Buffer D: 100 mM glycine, pH 2.5, adjusted with 1 M HCl. Store at 4°C.
9. Buffer E: same as Buffer B with 0.01% (w/v) sodium azide. Store at 4°C.

2.2. MALDI-TOF MS

1. MALDI-TOF instrument: Voyager DE-Pro (Applied Biosystems, Framingham, MA) or alternative mass spectrometer with delayed extraction and with a 337 nm nitrogen laser (see Note 2). Spectra are acquired in positive, linear acceleration mode, in the *m/z* 1,000–35,000 mass range. Settings: 24 kV of accelerating voltage, 76% of grid voltage and 180 ns of delay time. Averages of 10 × 50 laser shots are summed for one spectrum.
2. 1,2-Dimethoxy-4-hydroxycinnamic matrix (sinapinic acid, Sigma–Aldrich) is dissolved in 50% acetonitrile, 0.1% (v/v) trifluoroacetic acid solution in 10 mg/mL concentration. The matrix solution is prepared freshly. Matrix solution is light sensitive. Store at 20–22°C.

2.3. In-solution Enzymatic Digestion

1. Vivaspin 500 ultrafiltration devices with 3 kDa MWCO from Vivascience, Stonehouse, U.K. (see Note 3).
2. NH_4HCO_3. Stock solution is prepared at 100 mM concentration (see Note 4).
3. Tris[2-carboxyethyl]-phosphine-HC (TCEP) (Sigma–Aldrich). TCEP stock solution is prepared at 100 mM concentration in 100 mM NH_4HCO_3 (see Note 5). Prepare freshly.
4. Iodacetamide (IAA) (Sigma-Ultra, Sigma–Aldrich). IAA stock solution is prepared at 100 mM concentration in 100 mM NH_4HCO_3 (see Note 6). Prepare freshly before use, wrap the container with aluminum foil, and keep the solution in dark.
5. A trypsin stock solution is prepared at 100 ng/μL concentration in 5% acetic acid. Aliquots are stored at −20°C. Working solution is prepared freshly at 6 ng/μL concentration in 50 mM NH_4HCO_3. An appropriate volume of trypsin solution (2 μL) is added to the sample in order to achieve a 1:100 to 1:20 trypsin to protein ratio. Each working solution should be prepared freshly and kept on ice before use.
6. Protein LoBind Eppendorf tubes.
7. Water bath or heating block at 37°C and 100°C.

2.4. Nano-ESI-MS

1. ESI-MS instrument, QSTAR Elite (Applied Biosystems, Foster City, CA/Toronto, Canada) equipped with NanoSpray II ion source. Spectra are acquired in the positive ion mode using information-dependent analysis. Briefly, survey scans are performed in the range *m/z* 300–1,500 and the two most abundant multiply charged ions are automatically selected for MS/MS experiments. MS/MS scans are performed using nitrogen as collision gas in the range *m/z* 70–1,500 using dynamic collision energy setup. Ion source settings: 1,800 V capillary voltage, 200°C source heater, 60 V cone voltage.
2. Nano-HPLC, Ultimate 3000 Nano LC system (Dionex, Sunnyvale, CA, USA) equipped with two independent gradient pumps and an active nano flow-splitter. The instrument is connected to the ESI-MS instrument online and is set up for automated preconcentration and sample cleanup procedure using a two-position valve. The sample (10 μL) is preconcentrated and desalted by flushing solvent A through the trap column at 30 μL flow rate for 5 min. After cleanup, the two-position valve is switched to place the trap column in series with the separation column. Settings: 300 nL flow rate. Gradient: 5–50% B in 30 min 50–98% B in 6 s (see Note 7).
3. Nanoflow sprayer tip, home-made pulled silica capillary. Characteristics: 170 μm outer diameter, 100 μm inner diameter, tip 30 μm inner diameter (see Note 8).
4. Trap column, PepMap, C18, 300 μm inner diameter, 5 mm length, 300 Å pore size, and 5 μm particle size (LCPackings, Sunnyvale, CA, USA) or similar (see Note 9).
5. Separation column, PepMap, C18, 75 μm inner diameter 15 cm length, 300 Å pore size, and 3 μm particle size (LCPackings, Sunnyvale, CA, USA) or similar (see Note 10).
6. Solvent A: 2% acetonitrile in 0.1% formic acid and 0.025% trifluoroacetic acid. Solvents of highest available grade and water MilliQ or better should be used. Solvents are filtered and degassed.
7. Solvent B: 98% acetonitrile in 0.1% formic acid and 0.025% trifluoroacetic acid.

3. Methods

The method consists of four steps: (1) functionalization of the magnetic particles, (2) immunoaffinity capture of the toxin, (3) washing steps which may include elution of the toxin and the subsequent enzymatic digestion, and (4) MS detection (Fig. 1).

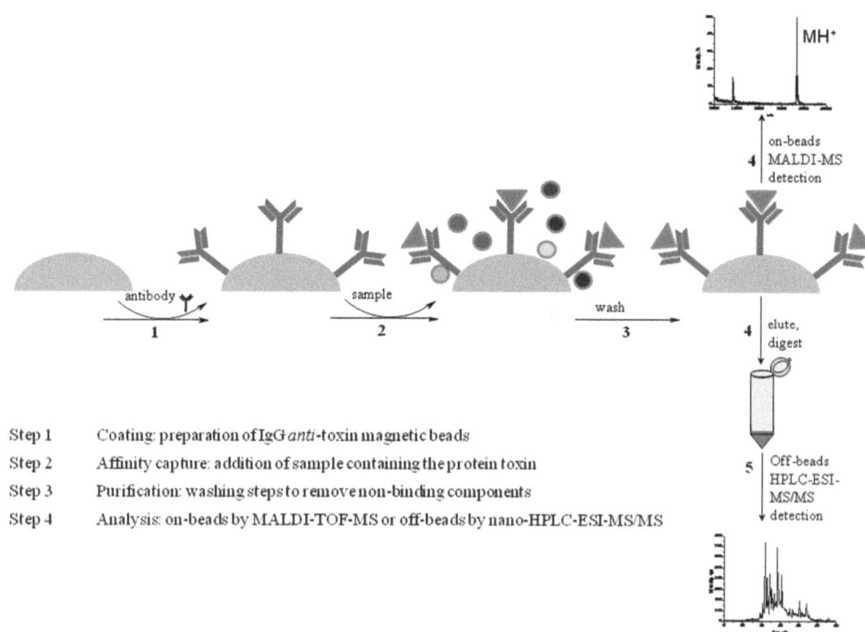

Fig. 1. Basic steps and schematics of solid phase magnetic immunocapture mass spectrometry. Symbols indicate the following: magnetic particles – *half circle*, antibody – *Y symbol*, antigen – *triangle* and impurities – *circles*.

Preparation of the affinity probe is performed by coating of monodisperse, hydrophobic magnetic particles with the antibody. It is important that the antibody is highly specific for the toxin antigen aimed to be analyzed. Antibody specificity should be evaluated by Western blot, ELISA, or MS-based pull-down assay using highly purified protein toxin antigen as standard. For the solid phase magnetic immunocapture assay, affinity-purified monoclonal antibodies are usually preferred to polyclonal ones. Antibodies can be covalently immobilized via reaction of their primary amino groups with the tosyl groups present on the magnetic beads surface. Loading capacity of the affinity probe has to be evaluated by ELISA using purified toxin standard.

In the immunoaffinity capture step, the sample is incubated and the bacterial protein toxin is selectively captured by the antibody-coated magnetic particles. After incubation, the particles are extensively washed in order to remove nonbinding molecules. The purified protein toxin bound to the particles is then either directly analyzed by matrix-assisted laser desorption/ionization time-of-flight mass spectrometry (MALDI-TOF-MS) (on-beads detection) or eluted with appropriate solvent, and subsequently off-beads detected. The sensitivity of MALDI-TOF-MS for intact proteins is in the low picomol–femtomol range, depending on molecular mass and primary structure of the intact protein.

Optimization of instrument parameters, data processing, and sample preparation procedures have to be performed for each analyte. Purified toxin standard should be used to determine detection limit (Fig. 2a). In the on-beads analysis, a suspension of the beads containing the affinity-captured protein toxin is placed

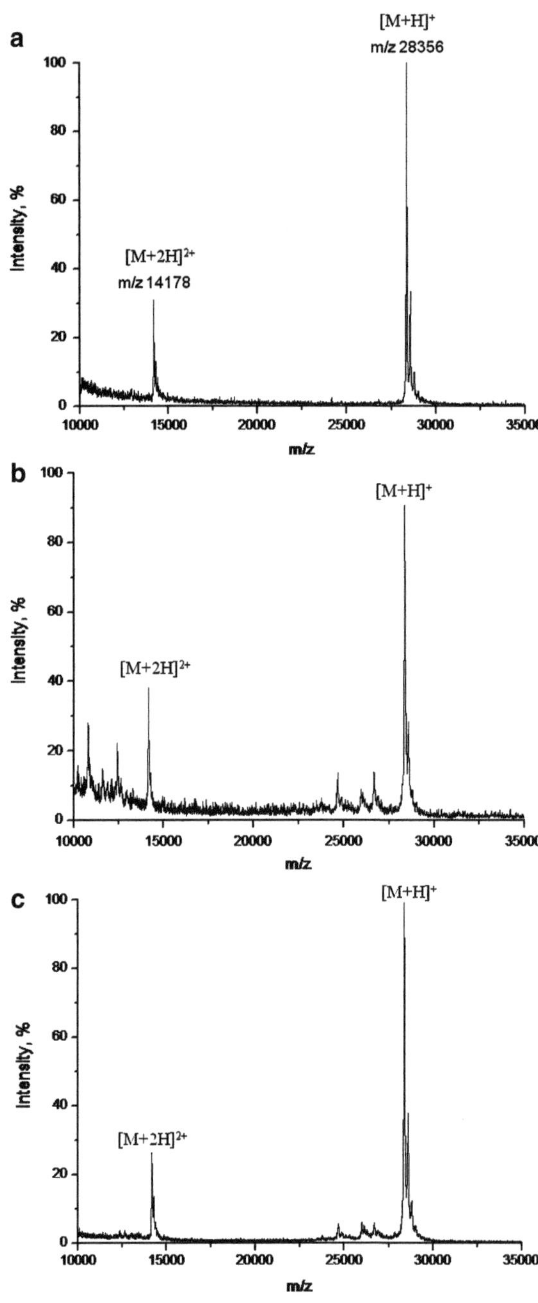

Fig. 2. MALDI-TOF mass spectrum of (**a**) SEB standard (85 fmol) and MALDI-TOF mass spectra of affinity-captured SEB in (**b**) on-beads and (**c**) off-beads detection modes.

directly onto the MALDI target plate, and the molecular mass of the protein toxin is measured in the linear mode (Fig. 2b). Magnetic beads were found to be highly compatible with MALDI under the experimental conditions typically applied for proteins. The acidic pH of the matrix (sinapinic acid) and the laser-induced desorption/ionization assist the dissociation of the antibody–antigen complex, and thus only the dissociated protein toxin can be detected in the mass spectrum. It can be easily and quickly performed by using up only small part (usually 1%) of the affinity probe. Alternatively, the protein toxin can be quickly eluted from the affinity probe by glycin at acidic pH. The presence of glycin (100 mM) in the toxin containing eluent was found not to influence negatively the subsequent MS-based analysis (Fig. 2c). The advantage of this elution step is that the affinity probe can be reused. In addition, being compatible not only with MALDI, but also with ESI, the elution step can increase the versatility of the assay. The eluted protein after pH adjustment can be proteolytically digested and analyzed by proteomics approaches which can give an additional dimension to the reliable identification and/or quantification of the toxin.

3.1. Preparation of Antibody-Coated Magnetic Particles and Immunomagnetic Isolation of Enterotoxins

1. 100 μL from the resuspended magnetic particles is washed twice with 400 μL Buffer A (see Note 11).
2. 60 μg antitoxin IgG (see Note 12) is diluted to 100 μL with Buffer A, added to the Dynabeads suspension and incubated under rotary shaking at 37°C for 24 h (see Note 13).
3. The supernatant is removed and particles are washed two times with 400 μL Buffer B at 4°C for 5 min.
4. The supernatant is removed, 400 μL Buffer C is added, and the suspension is incubated at 37°C for 4 h.
5. The supernatant is removed and the particles are washed two times with 400 μL Buffer B at 4°C for 5 min (see Note 14).
6. Particles are resuspended in the toxin-containing solution (sample) and incubated at 37°C for 1 h (see Note 15).
7. The supernatant is removed and the particles are washed five times with 400 μL Buffer B.
8. The supernatant is removed.

3.2. Online and Off-Line Detection of Bacterial Toxins Using MALDI-TOF MS

3.2.1. Online Detection

1. 1 μL suspension is loaded onto the MALDI target plate.
2. 1 μL matrix solution is added and the mixture is dried on air.
3. MALDI-TOF-MS analysis is performed. Bacterial protein toxin is identified based on the measured molecular mass.

3.2.2. Off-Line Detection

1. Particles are resuspended in 50 μL Buffer D.
2. The suspension is mixed for 1 min and the supernatant is removed (see Note 16).
3. 1 μL supernatant is mixed with 1 μL matrix solution, placed onto the MALDI target and dried on air.
4. MALDI-TOF-MS analysis is performed. Bacterial protein toxin is identified based on the measured molecular mass.

3.3. In-solution Trypsin Digestion and Off-Beads Nano-ESI-MS/MS Detection of Bacterial Toxins

1. Off-beads sample (supernatant obtained at 3.2.2.2) solution is exchanged from Buffer D to 100 mM NH_4HCO_3 using Vivaspin 500. The pH of the solution is checked after buffer exchange. In case of pH > 7.5, buffer exchange should be repeated.
2. The sample volume is adjusted to 40 μL with 100 mM NH_4HCO_3.
3. Add 5 μL TCEP stock solution and mix. Incubate the sample at 95°C for 10 min.
4. Cool down the sample to room temperature (20–22°C), add 5 μL IAA stock solution and shake in the dark for 20 min.
5. Add 2 μL working solution of trypsin, mix, and spin down.
6. Incubate for 4 h to overnight (18 h) at 37°C.
7. Acidify the sample to 2.5% final formic acid concentration using concentrated formic acid. Spin down the sample.
8. Nano-HPLC-ESI-MS/MS analysis is performed. Bacterial protein toxin is identified based on MS/MS spectra of tryptic peptide fragments.

4. Notes

1. SEs are produced by many isolates of *Staphylococcus aureus* (14). There are nine SEs (SEA–H) which have been characterized so far at the protein level, amongst which SEB is the most studied. SEB is known for causing significant nausea with vomiting, intestinal cramping, and diarrhea after hours of exposition. The toxic and lethal doses of SEB depend on the animal species and also on the route of exposure. Between 100 and 200 ng of ingested SEB can cause symptoms of staphylococcal intoxication. In humans, for aerosol exposures to SEB, currently the estimated 50% effective dose and 50% lethal dose are 0.4 and 20 ng/kg, respectively, for aerosol exposure (15).
2. Alternative similar benchtop MALDI-TOF mass spectrometers which can be used either in linear and reflectron modes are: microflex LRF (Bruker Daltonics) and M@ldi L/R (Waters, USA).

3. Vivaspin microcolumn is used according to manufacturer's instruction.
4. Prepare freshly and keep at 4°C.
5. TCEP is a valid alternative protein reduction agent to the commonly used dithiothreitol or 2-mercaptoethanol. TCEP is nonvolatile (odorless), more stable, and more effective than traditional reduction agents.
6. IAA is light-sensitive.
7. For HPLC, filtered (0.22 μm) and degassed solvents should be used. Solvents should be replaced weakly.
8. Similar nano-flow electrospray tips suitable for online ESI analysis are commercially available from New Objective, Woburn, MA USA.
9. Alternative commercial reverse phase C18 trap columns are (1) Symmetry 300 C18 NanoEase (Waters, USA) and (2) ZORBAX 300SB C18, 0.3×5 mm, 5-μm particles (Agilent Technologies, USA). Similar columns can be custom or homemade as well.
10. Alternative commercial reverse phase C18 nano-HPLC column is ZORBAX 300 SB C18, 75 μm×150 mm, 3.5-μm particles (Agilent Technologies, USA). Similar columns can be custom or home made as well.
11. High viscosity of the sample hinders the sedimentation of the magnetic particles.
12. Antitoxin IgGs are preferably affinity-purified and free of other proteins. These bind to the surface and reduce capacity of the magnetic particles.
13. All washing and incubation steps are performed under strong agitation to avoid sedimentation of the particles.
14. Antibody coated Dynabeads can be stored in Buffer D at 4°C for maximum 60 days. Before reuse, particles are washed five times with 400 μL Buffer B at 4°C for 5 min.
15. Volume of the sample is preferably below 1 mL.
16. Prolonged presence of antitoxin-coated particles in Buffer D causes loss of binding capacity.

References

1. Seto Y, Kanamori-Kataoka M (2005) Mass spectrometric strategy for the determination of natural and synthetic organic toxins. J Health Sci 51:519–525
2. Brun V, Dupuis A, Adrait A, Marcellin M, Thomas D, Court M, Vandenesch F, Garin J (2007) Isotope-labeled protein standards: toward absolute quantitative proteomics. Mol Cell Proteomics 6:2139–2149
3. Kalb SR, Moura H, Boyer AE, McWilliams LG, Pirkle JL, Barr JR (2006) The use of Endopep-MS for the detection of botulinum toxins A, B, E, and F in serum and stool samples. Anal Biochem 351:84–92

4. Hines HB, Lebeda F, Hale M, Brueggemann EE (2005) Characterization of botulinum progenitor toxins by mass spectrometry. Appl Environ Microbiol 71:4478–4486
5. Kalb SR, Moura H, Boyer AE, McWilliams LG, Pirkle JL, Barr JR (2006) The use of Endopep-MS for the detection of botulinum toxins A, B, E, and F in serum and stool samples. Anal Biochem 351:84–92
6. Kalb SR, Lou J, Garcia-Rodriguez C, Geren IN, Smith TJ, Moura H et al (2009) Extraction and inhibition of enzymatic activity of botulinum neurotoxins/A1, /A2, and /A3 by a panel of monoclonal anti-BoNT/A antibodies. PLoS One 4:5355
7. Boyer AE, Moura H, Woolfitt AR, Kalb SR, McWilliams LG, Pavlopoulos A et al (2005) From the mouse to the mass spectrometer: detection and differentiation of the endoproteinase activities of botulinum neurotoxins A-G by mass spectrometry. Anal Chem 77:3916–3924
8. Williams JP, Green BN, Smith DC, Jennings KR, Moore KAH, Slade SE et al (2005) Noncovalent Shiga-like toxin assemblies: characterization by means of mass spectrometry and tandem mass spectrometry. Biochemistry 44:8282–8290
9. van Baar BLM, Hulst AG, Roberts B, Wils ERJ (2002) Characterization of tetanus toxin, neat and in culture supernatant, by electrospray mass spectrometry. Anal Biochem 301:278–289
10. Boyer AE, Quinn CP, Woolfitt AR, Pirkle JL, McWilliams LG, Stamey KL et al (2007) Detection and quantification of anthrax lethal factor in serum by mass spectrometry. Anal Chem 79:8463–8470
11. Schlosser G, Kacer P, Kuzma M, Szilagyi Z, Sorrentino A, Manzo C et al (2007) Coupling immunomagnetic separation on magnetic beads with matrix-assisted laser desorption ionization-time of flight mass spectrometry for detection of staphylococcal enterotoxin B. Appl Environ Microbiol 73:6945–6952
12. Bernardo K, Fleer S, Pakulat N, Krut O, Hünger F, Krönke M (2002) Identification of *Staphylococcus aureus* exotoxins by combined sodium dodecyl sulfate gel electrophoresis and matrix-assisted laser desorption/ionization-time of flight mass spectrometry. Proteomics 2:740–746
13. Pauly D, Kirchner S, Stoermann B, Schreiber T, Kaulfuss S, Schade R et al (2009) Simultaneous quantification of five bacterial and plant toxins from complex matrices using a multiplexed fluorescent magnetic suspension assay. Analyst 134:2028–2039
14. Pocsfalvi G, Cacace G, Cuccurullo M, Serluca G, Sorrentino A, Schlosser G et al (2008) Proteomic analysis of exoproteins expressed by enterotoxigenic *Staphylococcus aureus* strains. Proteomics 8:2462–2476
15. Papageorgiou AC, Tranter HS, Acharya KR (1998) Crystal structure of microbial superantigen staphylococcal enterotoxin B at 1.5 A resolution: implications for superantigen recognition by MHC class II molecules and T-cell receptors. J Mol Biol 277:61–79

Chapter 2

Sensitive and Rapid Detection of Cholera Toxin-Producing *Vibrio cholerae* Using Loop-Mediated Isothermal Amplification

Wataru Yamazaki

Abstract

Loop-mediated isothermal amplification (LAMP) is an established nucleic acid amplification method offering rapid, accurate, and cost-effective diagnosis of infectious diseases. The LAMP assay requires 12–18 min for amplification with a single colony on selective agar from cholera toxin (CT)-producing *Vibrio cholerae* strains and less than 60 min with human feces and seafood samples. The assay requires less than 35 and 80 min for the detection of CT-producing *V. cholerae* with a colony on selective agar and with human feces and seafood samples from the beginning of DNA extraction to final determination. The LAMP amplification can be judged by both turbidimetric analysis and visual assessment with the unaided eye. The sensitivity of the LAMP assay is tenfold higher than that of the PCR assay. The LAMP assay is a powerful tool for rapid, simple, and sensitive detection of CT-producing *V. cholerae* which may facilitate the investigation of *V. cholerae* contamination in seafood, as well as the early diagnosis of cholera in humans.

Key words: Loop-mediated isothermal amplification, Cholera toxin, *Vibrio cholerae*, Rapid, simple, and sensitive detection, Human feces, Seafood

1. Introduction

Vibrio cholerae is widely acknowledged as one of the most important water- and seafood-borne pathogens causing outbreaks of diarrhea. Cholera toxin (CT) is a major virulence determinant of *V. cholerae*. These findings necessitate regular examination of *V. cholerae* isolates for their ability to produce CT in order to assess their clinical significance (1, 2). Detection of CT-producing *V. cholerae* using conventional culture-, biochemical-, and immunological-based assays is time-consuming and laborious,

Otto Holst (ed.), *Microbial Toxins: Methods and Protocols*, Methods in Molecular Biology, vol. 739,
DOI 10.1007/978-1-61779-102-4_2,

requiring more than 3 days. Although PCR assays provide more rapid identification of *V. cholerae* than conventional assays, they require the use of electrophoresis to detect amplified products, which is time-consuming and tedious (3, 4). Real-time PCR assays are not routinely used in remote or poor areas due to their requirement for an expensive thermal cycler.

Loop-mediated isothermal amplification (LAMP) is an established nucleic acid amplification method offering rapid, accurate, and cost-effective diagnosis of infectious diseases (5). The LAMP assay is faster and easier to perform than conventional PCR assays, as well as being more specific (5–7). This technique requires a simple heat block or a water bath providing a constant temperature (6). LAMP is based on the principle of autocycling strand displacement DNA synthesis performed by the *Bst* DNA polymerase large fragment for the detection of a specific DNA sequence with specific characteristics (5). This offers a number of advantages: first, all reactions can be carried out under isothermal conditions ranging from 60 to 65°C; second, its use of six primers recognizing eight distinct regions on the target nucleotides means that specificity is extremely high; and third, detection is simplified by visual assessment using the naked eye, without the need for electrophoresis.

The LAMP assay requires 12–18 min for amplification with a single colony on thiosulfate citrate bile salt sucrose (TCBS) agar or CHROMagar Vibrio from CT-producing *V. cholerae* strains and less than 45 min with spiked human feces (Fig. 1**a**) (8) and seafood samples. The assay requires less than 35 and 80 min for the detection of CT-producing *V. cholerae* with a colony on TCBS agar or CHROMagar Vibrio and with spiked human feces and seafood samples from the beginning of DNA extraction to final determination. Amplification in the LAMP assay can also be judged based on white precipitate by visual assessment using the naked eye (Fig. 1**b**). The sensitivity of the LAMP assay for CT-producing *V. cholerae* with spiked human feces is found to be 7.8×10^2 CFU/g (1.4 CFU/reaction) (Fig. 1**a**). The sensitivity of the LAMP assay is tenfold higher than that of the PCR assay (8). Its simplicity and rapidity are major benefits, since effective measures for outbreak containment depends on how fast the infectious agent is identified.

2. Materials

2.1. Bacterial Culture

1. Selective media, namely, TCBS agar (Eiken Chemical, Tokyo, Japan) and CHROMagar Vibrio (CHROMagar Microbiology, Paris, France).

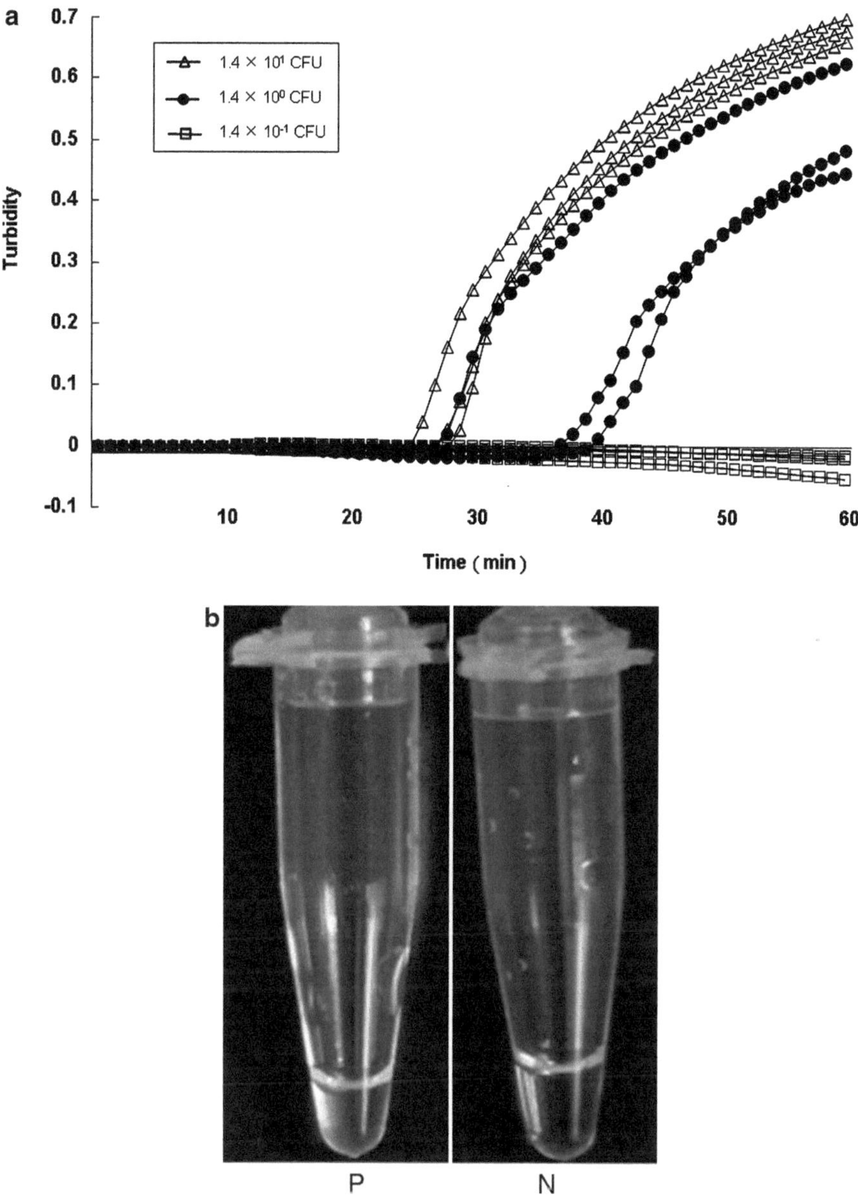

Fig. 1. Detection of CT-producing *V. cholerae* from spiked human feces by real-time turbidimeter and by observation of white precipitate. (**a**) Detection of CT-producing *V. cholerae* by real-time turbidimeter. The curves from *left* to *right* indicate decreasing concentrations of CFU from bacterial DNA (14.4 to 0.14 CFU/test tube) (Reproduced from ref. (8)). (**b**) Visual detection of CT-producing *V. cholerae* by observation of white precipitate. Tube P, positive sample; tube N, negative sample.

2.2. Human Fecal Samples

1. Fresh fecal samples obtained from clinical patients with suspected cholera infection.
2. Sterilized plastic tube (10 mL).
3. Phosphate buffered saline (PBS).

2.3. Seafood Samples

1. Fresh seafood samples with suspected contamination by *V. cholerae*.
2. Alkaline peptone water (APW, Eiken Chemical, Tokyo, Japan).
3. Stomacher (Pro-media, SH-001; ELMEX, Tokyo, Japan).
4. Sterilized plastic stomacher bag.

2.4. DNA Template

1. Heat block (for 95–100°C use).
2. Centrifuge for microcentrifuge tubes (900–20,000 × *g*).
3. Centrifuge for eight connected tubes.
4. Sterilized 1.5-mL microcentrifuge tube.
5. Sterilized 0.5-mL microcentrifuge tube.
6. 1 M NaOH stored at 20–22°C. Diluted 40-fold in sterile distilled water and adjusted to 25 mM, followed by storage at −20°C until use.
7. 1 M Tris–HCl buffer, pH 7.5, stored at 20–22°C.
8. Sterilized disposable loop (to be used for 1-μL inoculation).
9. Vortex mixer.

2.5. Primer Design and Preparation of Primer Mixture

Sequences and locations of each primer (synthesized by Hokkaido System Science Co., Ltd., Sapporo, Japan) are shown in Table 1 (see Note 1). The following primer volumes are required for each reaction (see Note 2).

0.4 μL of FIP (100 μM)
0.4 μL of BIP (100 μM)
0.2 μL of LF (100 μM)
0.2 μL of LB (100 μM)
0.05 μL of F3 (100 μM)
0.05 μL of B3 (100 μM)
(Total 1.3 μL)

The primer mixture was prepared in either 1.5 or 0.5-mL microcentrifuge tubes, followed by storage at −20°C until use.

2.6. LAMP Assay (see Note 3)

1. Loopamp DNA amplification kit (Eiken Chemical, Tokyo, Japan), which includes *Bst* DNA Polymerase, 2× Reaction Mix, and distilled water.
2. Heat block or water bath for endpoint detection, or Loopamp turbidimeter (Eiken Chemical, Tokyo, Japan) for both end-point and real-time detection (to be used at 65°C).
3. Loopamp LAMP reaction tube (Eiken Chemical, Tokyo, Japan).

Table 1
LAMP primers used

Primer	Sequence (5′ to 3′)	Gene location (bp)
CtxA-FIP	TCT GTC CTC TTG GCA TAA GAC GCA GAT TCT AGA CCT CCT G (F1c-F2)	277–257(F1c), 217–235(F2)
CtxA-BIP	TCA ACC TTT ATG ATC ATG CAA GAG GCT CAA ACT AAT TGA GGT GGA A (B1-B2c)	311–335(B1), 395–375(B2c)
CtxA-F3	GCA AAT GAT GAT AAG TTA TAT CGG (F3)	193–216
CtxA-B3	GMC CAG ACA ATA TAG TTT GAC C (B3c)	433–412
CtxA-LF	CAC CTG ACT GCT TTA TTT CA (LFc)	256–237
CtxA-LB	AAC TCA GAC GGG ATT TGT TAG G (LB)	336–357

Primer FIP consisted of the F1 complementary sequence and the F2 sequence. Primer BIP consisted of the B1 sequence and the B2 complementary sequence. Primer B3 and LF consisted of the B3 and LF complementary sequences, respectively

3. Methods

Although the LAMP assay is less affected by the inhibitory effects of clinical sample components than the PCR assay (9), the removal of inhibitory factors and concentration of the small number of target bacterial cells in clinical and food samples are essential for sensitive and reliable LAMP detection. Although commercially available kits, such as the QIAamp DNA Mini Kit (Qiagen, Hilden, Germany), offer a sophisticated approach to DNA extraction from clinical samples, these kits are time-consuming, laborious, and costly. To remove larger debris from fecal samples, food samples, and components of enrichment broths which contain DNA amplification inhibitors, as well as to concentrate the small number of bacterial cells, a simple, rapid, and cost-effective DNA extraction protocol was described using a combination of NaOH-heat treatment and three-step centrifugation procedures (8, 10).

During the DNA polymerization by *Bst* polymerase, a pyrophosphate ion is released from dNTP as a by-product. Production of large amounts of pyrophosphate ions lead to a reaction with magnesium ions from the LAMP reaction solution, which in turn produces magnesium pyrophosphate as simple turbidity. The increased turbidity in the reaction mixture caused by the production of insoluble white precipitate correlates with the amount of synthesized DNA (6). The white precipitate can be observed with the unaided eye, as well as measured for turbidity using a Loopamp turbidimeter.

3.1. DNA Extraction from Culture

1. Using a disposable loop (for 1 μL inoculation), inoculate a single loopful of fresh culture from selective media in a 1.5-mL microcentrifuge tube containing 50 μL of 25 mM NaOH (see Note 4).
2. Heat the cell mixture at 95–100°C for 5 min.
3. Add 4 μL of 1 M Tris–HCl buffer, pH 7.5, to neutralize the solution.
4. Centrifuge cell debris at 20,000 × *g*, 4°C for 5 min.
5. Use 1–2 μL of the supernatant as template DNA for the LAMP assay.

3.2. DNA Extraction from Human Fecal Sample

1. In a sterilized plastic tube (10 mL), prepare human fecal homogenates by adjusting the concentration to 10% using PBS.
2. Mix the homogenate using a vortex mixer, transfer 1 mL of the homogenate into a 1.5-mL microcentrifuge tube, followed by centrifugation at 900 × *g* for 1 min (see Note 5).
3. Transfer supernatant into a new 1.5-mL microcentrifuge tube.
4. Centrifuge for 5 min at 10,000 × *g*, and remove the supernatant (see Note 6).
5. Resuspend the pellets in 100 μL of 25 mM NaOH (see Note 4).
6. Mix the mixture using a vortex mixer, and then heat at 95–100°C for 5 min.
7. Add 8 μL of 1 M Tris–HCl buffer, pH 7.5, to neutralize the solution.
8. Centrifuge cell debris at 20,000 × *g*, 4°C for 5 min.
9. Use 1–2 μL of the supernatant as template DNA for the LAMP assay (see Note 7).

3.3. DNA Extraction from Seafood Sample

1. In a plastic stomacher bag, prepare seafood sample homogenates adjusted to a 10% concentration with APW in a plastic stomacher bag.
2. Treat the seafood/APW mixture from plastic stomacher bags by light hand massaging or by homogenization using a stomacher for 15–30 s (see Note 8).
3. Incubate the seafood/APW mixture at 35–37°C for 6–24 h.
4. Transfer 1 mL of the cultivated APW broth into a 1.5-mL microcentrifuge tube, and centrifuge at 900 × *g* for 1 min (see Note 5).
5. Transfer supernatant into a new 1.5-mL microcentrifuge tube.

6. Centrifuge for 5 min at 10,000 × *g*, and remove the supernatant (see Note 6).
7. Resuspend pellets in 100 μL of 25 mM NaOH (see Note 4).
8. Mix the mixture using a vortex mixer, and then heat at 95–100°C for 5 min.
9. Add 8 μL of 1 M Tris–HCl buffer, pH 7.5, to neutralize the solution.
10. Centrifuge cell debris at 20,000 × *g*, 4°C for 5 min.
11. Use 1–2 μL of the supernatant as template DNA for the LAMP assay.

3.4. Preparation of LAMP Reagents

1. Thaw LAMP reagents to 20–22°C, and then keep on ice.
2. Prepare the master mix in a 1.5 or 0.5-mL microcentrifuge tube while operating on ice. The following component amounts are required for each reaction.

   12.5 μL of 2× Reaction Mix.

   1.3 μL of primer mixture.

   1 μL of *Bst* DNA polymerase.

   8.2 μL of sterilized distilled water.
3. After dispensing, gently tap the tubes approximately two to three times.
4. Centrifuge the tubes for 2–3 s. The resulting mixture can be used as the master mix for the LAMP reaction.

3.5. Operation Procedure (Operate on Ice)

1. Dispense 23 μL of the master mix into Loopamp reaction tubes.
2. Add 1–2 μL of template DNA to the master mix, and adjust to a final volume of 24–25 μL.
3. Mix the mixture by pipetting or tapping (see Note 9).
4. Close the tube cap, and then centrifuge the tubes for 2–3 s (see Note 9).

3.6. LAMP reaction

1. Incubate the mixture at 65°C for 60 min using a Loopamp turbidimeter (endpoint or real-time detection), heat block, or water bath.
2. To terminate the reaction, inactivate the polymerase at 80°C for 5 min or 95°C for 2 min (see Note 10).
3. The reaction is considered positive when the turbidity reaches 0.1 within 60 min in a Loopamp turbidimeter, or when a white precipitate is visible to the unaided eye in the LAMP reaction tube using a Loopamp turbidimeter, heat block, and water bath (see Note 11).

4. Notes

1. All primers were designed from sequence data submitted to GenBank (Cholera toxin subunitA gene, *ctxA*, K02679) by Lockman et al. (11) with Primer ExplorerV4 software (http://primerexplorer.jp/elamp4.0.0/index.html; Fujitsu System Solutions, Tokyo, Japan). To find specific nucleotide sequences of CT-producing *V. cholerae*, a multiple alignment was determined with analyses of 34 *ctxA* sequences.
2. Use the highly purified LAMP primers for rapid and stably reproducible gene amplification. HPLC-grade purification is recommended for the production of FIP and BIP primers (see Table 1), whereas HPLC- or sequence-grade purification is required for the production of the other LAMP primers (LF, LB, F3, and B3, see Table 1).
3. The mechanism of the LAMP assay is complex and difficult to describe using simple diagrams. The Eiken Genome Web site clearly explains the details of the LAMP assay principle using a number of diagrams and animations (http://loopamp.eiken.co.jp/e/index.html). Given the high sensitivity of the LAMP assay in synthesizing large amounts of DNA, the amplification of the slightest amount of tainted product in the reaction may yield false-positive results. This type of contamination can be avoided by carrying out sample and reagent preparations on different clean benches. Amplification detection should be conducted using a turbidimeter, heat block, or water bath from which both reaction and detection can be accomplished while keeping the tube cap closed.
4. Using templates boiled using distilled water for DNA amplification may yield false-negative results, which can be avoided by NaOH-heat treatment, which potentially lyses and inactivates one or more unidentified inhibitory factors in bacterial cells during the LAMP reaction.
5. The first centrifugation for 1 min at $900 \times g$ is carried out to remove larger debris in the samples and enrichment broths, which in turn decreases the influence of inhibitory factors of fecal, seafood, and enrichment broth components.
6. The second centrifugation for 5 min at $10{,}000 \times g$ is carried out to concentrate the small number of bacterial cells in the samples. A total of 1 mL of 10% fecal homogenate and enrichment broth cultures are then concentrated to a volume of 108 and 54 μL by centrifugation, respectively.
7. Clinical patients excrete large amounts of *V. cholerae* cells in their feces. Therefore, enrichment procedures are not frequently performed to isolate this bacterium in fecal samples.

If required, add the fecal sample in a sterilized glass tube containing a ninefold APW broth, and follow the protocol for the detection of seafood sample, as described in Subheading 2.3.3.

8. Food components, such as organic and phenolic compounds, glycogen, fats, and calcium ions, were previously reported to inhibit DNA polymerase activity (12). Given that prolonged stomaching procedures appear to deteriorate the release of inhibitory factors from seafood samples, stomaching procedure should be performed within 30 s. Compared to the stomaching procedure, light hand massaging is preferable due to the low release of inhibitory factors (13).
9. Given that bubbles in the solution interfere with turbidity measurements and may cause false results, avoid creating bubbles when mixing the master mix and sample solutions. If bubbles are present, remove them by tapping the tubes approximately two to three times, and spin down the solution.
10. If the inactivation step at 80°C for 5 min or at 95°C for 2 min is omitted, visual assessment with the unaided eye should be performed quickly after the 60-min amplification. Failure to quickly assess the solution may lead to an observed nonspecific positive reaction in the negative sample due to residual heat in the LAMP reaction tube.
11. Caps from used LAMP reaction tubes should not be opened. Contamination of amplified LAMP products from other samples may lead to false interpretation of test results, as well as contaminate the testing area. Keep the caps of used tubes completely closed and dispose of tubes by incineration or after double bagging with a sealable vinyl bag. To prevent dispersion of the amplified LAMP products, do not perform autoclave sterilization treatment for disposal.

References

1. Faruque SM, Nair GB (2006) Epidemiology. In: Thompson FL, Austin B, Swings J (eds) The biology of *Vibrios*. ASM Press, Washington, DC, pp 385–398
2. Nishibuchi M, DePaola A (2005) *Vibrio* species. In: Fratamico PM, Bhunia AK, Smith JL (eds) Foodborne pathogens. Horizon Scientific, Norfolk, UK, pp 251–272
3. Shirai H, Nishibuchi M, Ramamurthy T, Bhattacharya SK, Pal SC, Takeda Y (1991) Polymerase chain reaction for detection of the cholera enterotoxin operon of *Vibrio cholerae*. J Clin Microbiol 29:2517–2521
4. Blackstone GM, Nordstrom JL, Bowen MD, Meyer RF, Imbro P, DePaola A (2007) Use of a real time PCR assay for detection of the *ctxA* gene of *Vibrio cholerae* in an environmental survey of Mobile Bay. J Microbiol Methods 68:254–259
5. Notomi T, Okayama H, Masubuchi H, Yonekawa T, Watanabe K, Amino N, Hase T (2000) Loop-mediated isothermal amplification of DNA. Nucleic Acids Res 28:E63
6. Mori Y, Nagamine K, Tomita N, Notomi T (2001) Detection of loop-mediated isothermal amplification reaction by turbidity derived from magnesium pyrophosphate formation. Biochem Biophys Res Commun 289:150–154

7. Nagamine K, Hase T, Notomi T (2002) Accelerated reaction by loop-mediated isothermal amplification using loop primers. Mol Cell Probes 16:223–229
8. Yamazaki W, Seto K, Taguchi M, Ishibashi M, Inoue K (2008) Sensitive and rapid detection of cholera toxin-producing *Vibrio cholerae* using a loop-mediated isothermal amplification. BMC Microbiol 8:94
9. Kaneko H, Kawana T, Fukushima E, Suzutani T (2007) Tolerance of loop-mediated isothermal amplification to a culture medium and biological substances. J Biochem Biophys Methods 70:499–501
10. Yamazaki W, Taguchi M, Kawai T, Kawatsu K, Sakata J, Inoue K, Misawa N (2009) Comparison of loop-mediated isothermal amplification assay and conventional culture methods for detection of *Campylobacter jejuni* and *Campylobacter coli* in naturally contaminated chicken meat samples. Appl Environ Microbiol 75:1597–1603
11. Lockman HA, Galen JE, Kaper JB (1984) *Vibrio cholerae* enterotoxin genes: nucleotide sequence analysis of DNA encoding ADP-ribosyltransferase. J Bacteriol 159: 1086–1089
12. Wilson IG (1997) Inhibition and facilitation of nucleic acid amplification. Appl Environ Microbiol 63:3741–3751
13. Kanki M, Sakata J, Taguchi M, Kumeda Y, Ishibashi M, Kawai T, Kawatsu K, Yamasaki W, Inoue K, Miyahara M (2009) Effect of sample preparation and bacterial concentration on *Salmonella enterica* detection in poultry meat using culture methods and PCR assaying of preenrichment broths. Food Microbiol 26:1–3

Chapter 3

Ultrasensitive Detection of Botulinum Neurotoxins and Anthrax Lethal Factor in Biological Samples by ALISSA

Karine Bagramyan and Markus Kalkum

Abstract

Both botulinum neurotoxins (BoNTs) and anthrax lethal factor, a component of anthrax toxin, exhibit zinc metalloprotease activity. The assay detailed here is capable of quantitatively detecting these proteins by measuring their enzymatic functions with high sensitivity. The detection method encompasses two steps: (1) specific target capture and enrichment and (2) cleavage of a fluorogenic substrate by the immobilized active target, the extent of which is quantitatively determined by differential fluorometry. Because a critical ingredient for the target enrichment is an immobilization matrix made out of hundreds of thousands of microscopic, antibody-coated beads, we have termed this detection method an *a*ssay with a *l*arge *i*mmuno-*s*orbent *s*urface *a*rea (ALISSA). The binding and reaction surface area in the ALISSA is approximately 30-fold larger than in most microtiter plate-based enzyme-linked immunosorbent assays (ELISAs). ALISSA reaches atto (10^{-18}) to femto (10^{-15}) molar sensitivities for the detection of BoNT serotypes A and E and anthrax lethal factor. In addition, ALISSA provides high specificity in complex biological matrices, such as serum and liquid foods, which may contain various other proteases and hydrolytic enzymes. This methodology can potentially be expanded to many other enzyme targets by selecting appropriate fluorogenic substrates and capture antibodies. Important requirements are that the enzyme remains active after being immobilized by the capture antibody and that the substrate is specifically converted by the immobilized enzyme target at a fast conversion rate.

A detailed protocol to conduct ALISSA for the detection and quantification of BoNT serotypes A and E and anthrax lethal factor is described.

Key words: Biological warfare, Botulinum neurotoxin, Anthrax lethal factor, Detection, Quantification, Fluorogenic peptide, Immunosorbent matrix, Protein A/G beads

1. Introduction

Botulinum neurotoxins (BoNTs) are considered the most potent toxins known. By extrapolation from primate studies, the lethal human dose is 1–2 ng/kg body weight when intravenously injected (1). Seven BoNT serotypes (A–G) are known to be produced by

Otto Holst (ed.), *Microbial Toxins: Methods and Protocols*, Methods in Molecular Biology, vol. 739,
DOI 10.1007/978-1-61779-102-4_3,

Gram-positive anaerobic bacteria of the genus *Clostridium* (2). BoNT/A and B (and to some extent E and F) are the main etiological agents of human botulism (3). Infant, food-borne and wound botulism are its most common forms (4–6). Natural BoNT is produced as a 900-kDa complex that contains the 150-kDa holotoxin consisting of a 50-kDa light and 100-kDa heavy chain, plus several nontoxic neurotoxin-associated proteins (NAPs) (7, 8). Once in the bloodstream, the BoNT holotoxin targets and enters motor neurons, inside which the toxin's light chain zinc metalloprotease subunit hydrolyzes SNARE proteins (9). BoNT-cleaved SNARE proteins no longer mediate the fusion of acetylcholine-containing synaptic vesicles with the terminal motor neuron membrane (10). This efficiently shuts down neurotransmitter release into the neuromuscular junction, leading to flaccid paralysis on the macroscopic scale. Each BoNT serotype cleaves one or more of the three SNARE proteins (SNAP25, VAMP, and syntaxin) at specific peptide bonds (10–14).

BoNTs have gained popularity as cosmetic drugs in recent years, and have also been successfully used for the treatment of a variety of neurological and neuromuscular disorders (15, 16). However, because of the lack of a standardized testing procedure, the units of biological activity are often unable to be directly converted into precise doses for human use, and overtreatment with BoNTs can cause iatrogenic forms of botulism (17, 18). BoNT is also a potential biothreat agent because of its extreme potency and lethality, its ease of production and transport, and the need for prolonged intensive care of intoxicated persons (2).

The clinical diagnosis of botulism requires the presence of the toxin be demonstrated in a clinical specimen. The mouse bioassay is most commonly used. For example, it is applied for the analysis of stool and enema samples from suspected cases of infant botulism (19, 20). Mice are intraperitoneally injected with a sterile-filtered sample and observed for signs of botulism. Furthermore, neutralizing antibodies can be used to specify the serotype of the causative BoNT. The mouse bioassay has a detection limit of 10–20 pg of neurotoxin, and typically requires up to 4 days turnaround time (19). The assay detailed here can detect BoNT/A and BoNT/E with atto (10^{-18}) and femto (10^{-15}) molar sensitivity, respectively, and requires only a fraction of the time of the mouse bioassay (2.5 h ALISSA for BoNT/A) (21).

Another zinc metalloprotease for which we have successfully adapted the same type of assay method is anthrax lethal factor (LF). LF constitutes one of the three components of anthrax toxin that is produced by *Bacillus anthracis*, together with protective antigen (PA) and edema factor (EF) (22). LF specifically cleaves members of the mitogen-activated protein kinase kinase (MAPKK) family, leading to the inhibition of essential signaling pathways. LF alone is not toxic; it requires the presence of PA for its translocation into cells (22). Macrophages are believed to be primarily affected by LF (23). A specific and sensitive assay for the detection of LF is potentially useful for early diagnosis of anthrax infection and is expected to be a useful research tool to advance the understanding of the mechanism of action of anthrax toxin (24).

1.1. Detection Principle

We refer to the method described here as an *a*ssay with a *l*arge *i*mmuno-*s*orbent *s*urface *a*rea (ALISSA) (21). ALISSA measures the specific proteolytic activity of BoNT/A, BoNT/E, or LF after its capture and enrichment on a beaded immunoaffinity matrix that contains target-specific antibodies. Antibodies were chosen such that the enzyme retains its catalytic activity after immobilization. Following the removal of nontoxin-specific sample components through stringent washes, the toxin-specific enzyme activities are determined by measuring the cleavage of specific fluorogenic peptide substrates (21). Fluorogenic peptide substrates for BoNTs and LF are commercially available, or can be synthesized by classical solid-phase peptide chemistry (25). The fluorogenic substrates contain a fluorescence donor and acceptor pair. Fluorescence of the fluorophore is quenched via the Förster resonance energy transfer (FRET) effect when the donor and acceptor are in close proximity to each other (preferably less than 10 nm) (26). The toxin's proteolytic activity hydrolyzes a peptide bond that leads to the separation of donor and acceptor, thereby releasing the unquenched fluorescence of the fluorophore (Fig. 1). In contrast to the classical FRET effect, the acceptors

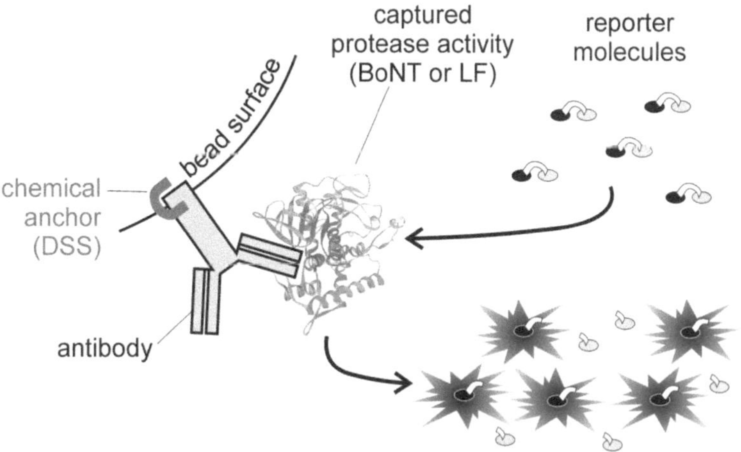

Fig. 1. Principle of the ALISSA. Antibodies against targets, such as BoNTs or LF, are conjugated on protein A-coated beads via their Fc regions and then cross-linked. An immobilized toxin protease molecule cleaves fluorogenic reporter molecules and releases unquenched fluorescent products.

used here are nonfluorescent. Therefore, the difference in the fluorescence of the donor before and after enzymatic cleavage is measured in the ALISSA. The peptide substrates used here contain N-terminal fluorescein labels, such as 5-carboxyfluorescein (5Fam) or fluorescein isothiocyanate (FITC) as donors with 4,4-dimethylamino-azobenzene-4′-carboxylic acid (Dabcyl) as an acceptor near the C-terminus; or alternatively, they contain a donor/acceptor pair of ortho-aminobenzoic acid (*o*-Abz)/2,4-dinitrophenyl (dnp).

Another critical component of the ALISSA is the beaded immunomatrix. Its immunosorbent surface area is approximately 30-fold larger than the typical microplate well surface of a classical enzyme-linked immunosorbent assay (ELISA), and at least 5 μg of antibody are used for each data point of an ALISSA measurement (21). The target-binding light chains of the antibodies are directed away from the bead surface and toward the sample solution by immobilization via their heavy chains, the Fc regions, on protein A/G-coated agarose beads. Bleeding of antibodies into the sample solution would have a detrimental effect on the assay's sensitivity. Therefore, the antibodies are loaded well below the specified loading capacity of the beads and are covalently conjugated with the beads by an irreversible chemical cross-linker.

The ALISSA not only concentrates and preserves enzyme activities of the immobilized toxin, but also has a profound effect on kinetic properties. Enzymatic turnover rates of ALISSA-immobilized enzymes are dramatically increased over those of the reaction of nonimmobilized enzymes, leading to a strong signal amplification effect on the order of billions of cleaved substrate molecules per captured toxin molecule per hour (21).

We have originally described the ALISSA for BoNT/A in spiked samples of human serum, gelatin phosphate diluent (GPD, used in clinical diagnosis of infant botulism) and liquid foods, such as milk and carrot juice (21). ALISSA's high sensitivity (attomolar detection of BoNT/A) is accompanied by high target specificity and robustness. Nontoxin proteases of most sample types do not lead to false-positive ALISSA signals because they are removed in stringent wash steps (Fig. 2).

ALISSA has now been expanded to detect anthrax LF and BoNT serotype E, but the detection of many other targets seems feasible, provided that suitable substrates and antibodies can be obtained. Thus, ALISSA has the potential to significantly improve the diagnosis of botulism, anthrax infection and potentially other serious infections, and could serve to protect humans in biomedical and biodefense scenarios.

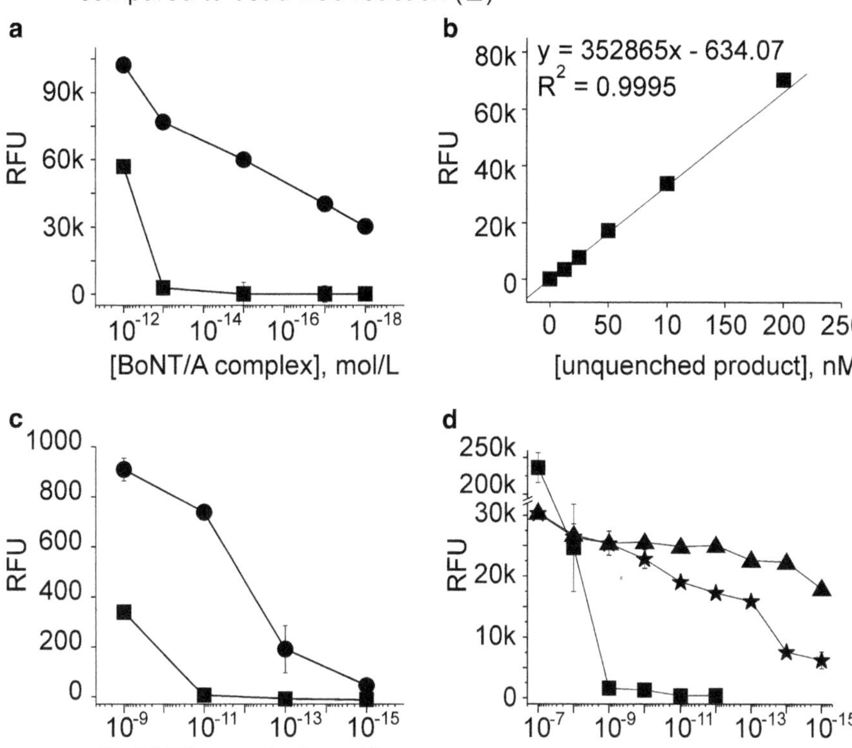

Fig. 2. ALISSA examples for BoNT/A, E, and anthrax LF. Results of the bead-free reaction (substrate and toxin only) are compared to the bead-based ALISSA (*see* components) for dilution series of the toxins in serum (**a**, **c**, **d**). ALISSA components were: rabbit polyclonal antibodies to *Clostridium botulinum* A toxoid, protein A/G beads, fluorogenic peptide (SNAPtide), and BoNT/A complex (**a**); Rabbit polyclonal antibodies against BoNT/E, protein A/G beads, fluorogenic substrate (SNAP Etide), and BoNT/E complex (**c**); ALISSA with: goat anti-anthrax LF, protein A and G beads and peptide substrate (MAPKKide), and LF (**d**); Standard curve of the fluorescence signal of unquenched calibration peptide (SNAPtide) in BoNT/A reaction buffer (**b**).

2. Materials

2.1. Instruments

1. 100SD Microcentrifuge (USA Scientific, USA); for more than five assays use a microcentrifuge, such as Eppendorf Model 5417R, or a centrifuge that accommodates 15-mL conical tubes (e.g., Beckman Allegra 6R).
2. Microplate mixer (e.g., Multi-Microplate Genie, Scientific Industries, Inc., USA).
3. Rotisserie: Labquake (Barnstead International, Dubuque, IA).
4. Microtiter plate reading spectrofluorometer: e.g., Wallac 1420 Multilabel Counter Victor2 (PerkinElmer) or SpectraMax M2 or better (Molecular Devices, USA) (see Note 1).
5. Standard laboratory incubator (37°C).

2.2. Plastic Ware (See Note 2)

1. Columns: Mobicol 5 mL and 1 mL columns with 10-μm pore size for lower and upper filters (MoBiTec GmbH, Germany) or Screw cap spin columns (Pierce, USA) (see Note 3).
2. Microcentrifuge collection tubes, 2 mL.
3. Microcentrifuge sample tubes, 1.5 mL.
4. Conical polypropylene tubes, 15 mL (BD, Falcon, USA).
5. Seal-Rite 2.0 mL microcentrifuge tube, amber (USA Scientific).
6. Seal-Rite 0.5 mL microcentrifuge tube, amber (USA Scientific).
7. 96-well 300 or 150-μL black microplates or 96-well 150 μL black microplates (Whatman or Greiner, USA).
8. Luer-Lock disposable syringes: 1, 3, and 5 mL (Cole-Parmer, USA).

2.3. Reagents

2.3.1. Botulinum Neurotoxins (See Note 4)

1. BoNT from *Clostridium botulinum* serotypes A and E in any of these forms: 900-kDa complex, 150-kDa holotoxin, or 50-kDa light chain (List Biological Laboratories, USA or Metabiologics, USA).

2.3.2. Anthrax Lethal Factor

1. Anthrax lethal factor, recombinant from *B. anthracis* (List Biological Laboratories, USA).

2.3.3. Antibodies

1. Rabbit polyclonal antibody for *C. botulinum* A Toxoid, ab20641 (Abcam, Cambridge, MA) or mouse monoclonal antibody for light chain of BoNT/A (F1-40, gift from Dr. Larry Stanker, US Department of Agriculture, USA).
2. Rabbit polyclonal antibody for *C. botulinum* toxin serotypes E (Metabiologics, Inc., Madison, WI).
3. Goat anti-lethal factor from *B. anthracis* (List Biological Laboratories, USA).

2.3.4. ALISSA Substrates (See Note 5)

1. SNAPtide peptide substrate (FITC/Dabcyl) for *C. botulinum* neurotoxin type A, or SNAPtide (*o*-Abz/Dnp) peptide substrate for neurotoxin type A (List Biological Laboratories, USA).
2. SNAPtide, unquenched calibration peptides for SNAPtide peptide substrate with FITC/Dabcyl and *o*-Abz/Dnp (List Biological Laboratories, USA).
3. SNAP Etide (*o*-Abz/Dnp) peptide substrate for *C. botulinum* neurotoxin type E (List Biological Laboratories, USA).
4. MAPKKide peptide substrate (Dabcyl/FITC) for anthrax lethal factor (List Biological Laboratories, USA).
5. As an alternative to FITC-labeled SNAPtide, it is possible to obtain custom-synthesized peptides from various sources. We successfully used a fluorogenic BoNT/A substrate with

the following sequence: 5Fam-TRIDEANQRATK(Dabcyl) X-amide, where 5Fam is at the α-amino group, Dabcyl at the ε-amino group of lysine, and X is norleucine with an amide C-terminus. The 5Fam label is much more stable than the FITC-labeled N-terminus, leading to lower background fluorescence.

2.3.5. Reagents and Buffer Solutions

1. Deionized water (ultrapure) with 18 MΩ/cm or lower conductivity, protease-free, 0.2 μm filtered, and autoclaved.
2. 4-(2-Hydroxyethyl)-1-piperazineethanesulfonic acid (HEPES acid).
3. Acetonitrile, CH_3CN (30%).
4. Zinc chloride, $ZnCl_2$.
5. Sodium chloride, NaCl.
6. Tween-20.
7. Dimethyl sulfoxide (DMSO).
8. Ethylenediaminetetraacetic acid (EDTA), 20 mM, pH 8.2.
9. Disuccinimidyl suberate (DSS), No-Weigh format, M.W. 368.35 g/mol, spacer arm 11.4 Å, 8×2 mg vials (Pierce, USA).
10. Pooled human serum (Sigma or Innovative Research, USA).
11. HEPES potassium salt.
12. Immobilized protein A/G Plus, 50% slurry (Pierce, USA) or immobilized protein A/G Plus, 25% slurry (Santa Cruz Biotechnology, Inc., USA).
13. *Coupling buffer (10×):* 100 mM sodium phosphate, 1.5 M NaCl, pH 7.2.
14. *Immunoprecipitation (IP)/wash buffer*: 25 mM Tris, 150 mM NaCl, 1 mM EDTA, 1% Nonidet P-40 (NP-40, US Biological, USA), 5% glycerol, pH 7.4.
15. *Conditioning buffer (100×)*: neutral pH buffer (Pierce Crosslink Immunoprecipitation Kit, Pierce, USA).
16. *Elution buffer:* pH 2.8 (Crosslink Immunoprecipitation Kit, Pierce, USA), or 100 mM glycine HCl buffer, pH 2.8, or from Gentle Ag/Ab Binding and Elution Buffer Kit (Pierce, USA).
17. *Reconstitution and reaction buffer for BoNT/A light chain:* 20 mM HEPES, pH 8.2, 0.5 mg/mL bovine serum albumin (BSA) or 0.1% Tween-20.
18. *Reconstitution buffer for BoNT/A holotoxin:* 20 mM HEPES, pH 8.0, 2.5 mM DTT, 0.3 mM $ZnCl_2$, and 1.0 mg/mL BSA.
19. *Reaction buffer for BoNT/A holotoxin and complex:* 20 mM HEPES, 0.3 mM, $ZnCl_2$, 2.5 mM DTT, and 0.1% Tween-20, pH 8.0.

20. *Reconstitution buffer for BoNT/E:* 50 mM HEPES, pH 7.8, 0.1% Tween-20.
21. *Reaction buffer for BoNT/E:* 50 mM HEPES, pH 7.8, 200 mM NaCl, 0.1% Tween-20.
22. *Reconstitution and reaction buffer for anthrax LF:* 20 mM HEPES, pH 7.2, 125 μg/mL BSA.
23. *Stock solutions for fluorogenic peptides*: The quenched SNAPtides, SNAP Etide, and MAPKKide are prepared as 2.5 mM stock solutions and the unquenched calibration peptides as 1 μM stocks, all in DMSO (see Note 5).

3. Methods

3.1. Preparation of the Immunomatrix

3.1.1. Binding of Antibody to Protein A/G Plus Agarose

The following protocol was adapted from the *Pierce Crosslink Immunoprecipitation* (IP) protocol (Pierce, Rockford, IL, USA) with some modifications. This protocol is designed to yield enough antibody-coated beads for ten assay reactions (5 μg antibody per reaction and data point). We recommend conducting the ALISSA at least in duplicate (two reactions per sample). The procedure can be proportionally scaled to prepare immunomatrix for up to 20 assay reactions using the same plastic ware and column sizes.

1. Prepare 10 mL of coupling buffer (1×) for each IP reaction by diluting the coupling buffer (10×) with ultrapure water.
2. Evenly suspend the protein A/G plus agarose beads by swirling the bottle gently. Using a pipettor equipped with a cut pipette tip (see Note 6) add 100 μL of the suspended Pierce resin slurry or, alternatively, 200 μL of the Santa Cruz resin slurry per reaction to a 5-mL Mobicol column.
3. Place column into a 15-mL conical polypropylene tube. Attach the Luer Lock cap to the Mobicol column and connect to a 5-mL air-filled disposable syringe. Gently press the plunger into the syringe until all liquid has left the column. Alternatively, centrifuge (see Note 7). Discard the flow-through.
4. Wash the resin in the column by adding 1.0 mL of coupling buffer (1×) and discard the flow-through using the syringe technique described in step 3. Repeat this wash once.
5. In a separate microcentrifuge tube mix coupling buffer (10×), water, and 50 μg of antibody in solution to yield a final volume of 500 μL at a dilution of coupling buffer (1×) (see Note 8). For example, if the antibody is received at a concentration of 1 μg/μL, mix 50 μL of coupling buffer (10×) with 400 μL water and then combine with 50 μL of the antibody solution.

6. Gently tap the bottom of the column on a paper towel to remove any excess liquid. Insert the bottom plug to block the column's drain.
7. Immediately transfer the antibody solution prepared in step 5 into the resin-containing column. Never allow the beads to dry out.
8. Attach the closing screw cap to the column and incubate on the rotating rotisserie at 22°C for 1 h. Ensure that the slurry remains suspended at all times during incubation.
9. Remove and save bottom plug and cap. Place the column into a 15-mL conical collection tube and remove the liquid as done in step 3. Save the flow-through to verify antibody coupling by Bradford or bicinchoninic acid (BCA) protein assay (see Note 9).
10. Wash the resin once with 0.5 mL and then twice with 1.5 mL of coupling buffer (1×), remove and discard the flow-through as done in step 3.

3.1.2. Cross-linking of the Protein A/G-Coupled Antibody (See Note 8)

1. To prepare a fresh 25-mM DSS solution, add 217 μL DMSO into a new vial of 2 mg DSS, by inserting the pipette tip through the vial's foil covering. Thoroughly mix the solution by multiple pipette aspiration and dispense steps until all DSS is fully dissolved.
2. In a new microfuge tube, further dilute the 25 mM DSS solution 1:10 with DMSO to yield 2.5 mM DSS.
3. Tap the bottom of the column with antibody-coated resin from 4.1.1 on a paper towel to remove excess liquid and insert the bottom plug.
4. Add 205 μL coupling buffer (1×) to the antibody-coated resin and resuspend. Then, combine with 45 μL of 2.5 mM DSS solution (step 2) to obtain a concentration of 450 μM DSS. Cap the column with the closing screw cap.
5. Incubate the cross-linking reaction for 1 h at 22°C on the rotisserie.
6. Remove and save bottom plug and cap. Place the column into a collection tube and drain the liquid as done in Subheading 3.1.1, step 3.
7. Add 250 μL of elution buffer to the column and drain the liquid gently, as done above. Save the flow-through to verify antibody cross-linking by measuring protein concentration (as in Subheading 3.1.1, step 9).
8. Wash twice with 0.5 mL of elution buffer to remove non-cross-linked antibody and unreacted DSS.
9. Wash twice with 1.0 mL of ice-cold IP/wash buffer and drain the liquid after each wash.

10. Add 0.5 mL of IP/wash buffer and transfer the resin into a new tube. Fifty microliter of this resin suspension contains beads with approximately 5 μg antibody.

3.1.3. Storage Conditions

The resin with the cross-linked antibody can be stored for up to 5 days in IP/wash buffer at 4°C. It is recommended to transfer antibody-bound resin into a new, protease-free microcentrifuge tube. For longer storage (maximum of 2 weeks at 4°C), store resin in coupling buffer (1×) and wrap the tube with parafilm.

3.2. ALISSA for BoNT and LF

The protocol below describes the amounts and volumes used for a single ALISSA reaction (one data point). Each reaction is conducted in a separate tube. Scale the number of tubes according to the number of samples to be analyzed.

3.2.1. Toxin Enrichment and Immobilization

1. Combine the sample to be analyzed (e.g., serum and liquid food) with IP/wash buffer in a 3:1 ratio. The final volume of sample mixed with IP/wash buffer may range from 0.5 to 5 mL. Use 15-mL conical sterile tubes for sample volumes larger than 1 mL and microcentrifuge tubes for samples less than 1 mL.
2. Add 50 μL of resin with cross-linked antibody from Subheading 3.1.2, step 10, to the sample/IP/wash buffer mix and incubate by gentle mixing on the rotisserie for 1–2 h at 22°C or 16 h at 4°C.
3. Use a syringe to transfer the resin-containing sample into a spin cup column (Mobicol 1 mL or Pierce 1 mL spin column) with Luer-Lock connector and place into a collection tube (Fig. 3a, see Note 10). Discard the liquid.
4. Add 200 μL of IP/wash buffer to the resin and centrifuge with spin column placed into collection tube (Fig. 3b). Discard flow-through. As an alternative wash buffer, 20 mM HEPES, pH 7.5, with 0.1% Tween-20, can be used for BoNT assays. Repeat five times.
5. Wash resin once with 100 μL of conditioning buffer (1×).
6. Wash resin once with 100 μL of 2.5 M NaCl.
7. Wash resin twice with 200 μL of ultrapure protease-free water.
8. Resuspend resin in 100 μL of ultrapure protease-free water, and transfer into a new microcentrifuge tube using a cut pipette tip (see Note 6). Wrap with parafilm and store at 4°C until used (up to 5 days).

3.2.2. Reaction of the Fluorogenic Peptide Substrate

1. The final concentration of the fluorogenic substrate in the reaction buffer should be 5 μM. Therefore, the peptide stock solutions need to be diluted. Prepare 250 μM prediluted stock of the appropriate fluorogenic peptide substrate by

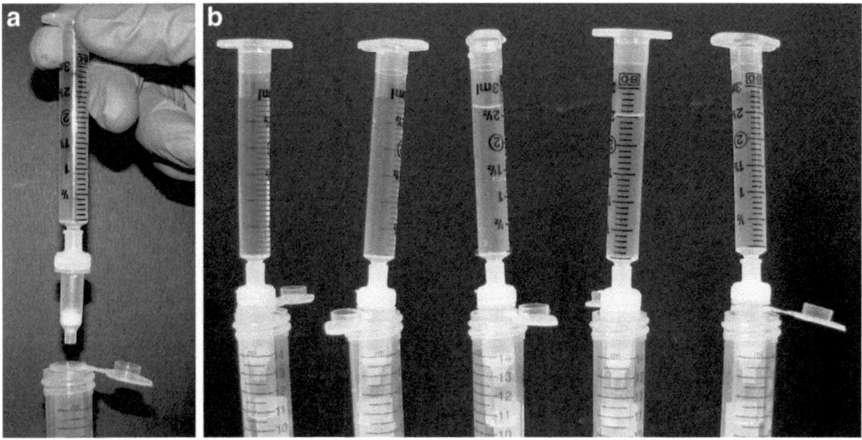

Fig. 3. Handling of the beaded ALISSA resin in spin columns. Loading of a spin column after the immunoprecipitation step using a Luer-Lock adaptor screw cap (**a**). Resins wash with multiple samples by gravity flow (**b**).

diluting the 2.5 mM peptide stock solution with 30% CH_3CN in ultrapure protease-free water. Each reaction requires 10 μL of the prediluted 250-μM substrate stock.

2. Prepare a reaction buffer by diluting the prediluted fluorogenic peptide stock 50-fold in the appropriate toxin reaction buffer (defined in Subheading 2.3.5.). Each sample requires 450 μL of reaction buffer.
3. In a 2.0-mL amber microcentrifuge tube, combine 50 μL of resin of the immobilized toxin sample with 450 μL of the substrate-containing reaction buffer. As a control, include a sample of 450 μL substrate-containing reaction buffer with 50 μL toxin-free resin. This control is used to establish the baseline of the fluorescence background.
4. Incubate 1–3 h by gently rotating the amber reaction tube(s) on the rotisserie inside an incubator at 37°C (see Note 11).
5. Transfer 220 μL of the reaction mixture (beads included) into a well of a 96-well black microplate (300 μL well volume).
6. Measure fluorescence (see Note 12) with a Wallac 1420 Multilabel Counter Victor2 spectrofluorometer or comparable plate reader. Subtract the fluorescence intensity of the baseline (toxin-free control, see step 3) from the fluorescence intensity of each of the measured samples.

3.2.3. Calibration Curve

The following protocol was adapted from the List Biological Laboratories Standard Curve protocol (Campbell, CA, USA) with some modifications.

1. Use calibration peptides that correspond to the cleaved, fluorescent substrate product to generate a calibration curve that allows conversion of relative fluorescence units (RFU) into

Table 1
Dilution table for standard curve

Final concentration of calibration peptide (μM)	Volume of 1.0 μM calibration peptide (μL)	Volume of reaction buffer (μL)
0.5000	500	500
0.2000	200	800
0.1000	100	900
0.0500	50	950
0.0250	25	975
0.0125	10	990
0.0000	0	220

molar units of cleaved substrate. This value can then be used to calculate specific enzymatic activity, when the reaction time, volume, and protein amount are also recorded (21).

2. Prepare a 0.5-mM stock solution by dissolving 1 vial of calibration peptide (~49.4 nmol) in 98.8 μL of DMSO.
3. Further dilute the calibration peptide to a 1.0-μM solution by adding 5 μL of the 0.5 mM calibration peptide stock solution to 2,495 μL of toxin reaction buffer.
4. Each dilution is done in triplicate using 220 μL/well for a 300-μL well volume (96-well black microplate) or 120 μL of total reaction buffer for a 150-μL well volume (reduced volume 96-well black microplate). Prepare the dilution series (see Table 1).

4. Notes

1. In our hands, the Victor2 spectrofluorometer performed at much higher sensitivity than the SpectraMax M2 instrument. However, the SpectraMax M2 can be manually set to any desired excitation/emission wavelength in the UV/VIS spectrum, which is very helpful for initial assay design. In contrast, the Victor2 requires a set of optical filters with fixed transmission wavelengths.
2. All plastic ware must be sterile and protease-free; preferably autoclaved.
3. Spin columns can be recycled for reuse in the following way: Remove the upper filter and wash the column multiple (at least five) times in 70% ethanol and then in ultrapure

protease-free water (ten times), dry them, and equip with a new upper filter (MoBiTec, Germany).

4. In some countries, the possession and laboratory use of BoNTs is regulated by law, and restrictions regarding permissible amounts and shipping exist. The catalytic BoNT light chain and anthrax LF are generally accepted as nontoxic. BoNTs should be handled with extreme caution and aerosol formation must be avoided. The use of a class II or III biological safety cabinet is recommended for experimentation with BoNTs.
5. Stock solutions of reconstituted fluorogenic substrates can be kept at −20°C in small aliquots. Avoid unnecessary freeze/thaw cycles. Dry peptide powders should be stored in a desiccator in the cold. All fluorogenic substrates should be kept in the dark.
6. Cut approximately 0.5 mm off the plastic tip of a 300-μL Rainin pipette using a sterile single-sided razor blade or scalpel.
7. Perform all resin centrifugation for 1 min at approximately 1,000 × *g* at 22°C.
8. This protocol is optimized for 5 μg of antibody per enzymatic reaction (one data point). Depending on the number of assays, proportionally scale the amount of antibody, resin, cross-linker, and buffer volumes.
9. Bradford protein quantification assay (Bio Rad, USA) or BCA protein assay (Pierce, USA).
10. Sample volumes larger than 1 mL may also be processed using 5-mL Mobicol columns with filter inserts. If done so, it is recommended to perform the liquid removal by centrifugation rather than with syringes.
11. *o*-Abz-conjugated peptides react slowly, therefore, the reaction may be prolonged for up to 16 h.
12. For BoNT and LF reactions with FITC/Dabcyl-peptides, use $\lambda_{ex} = 485$ nm and $\lambda_{em} = 535$ nm and for *o*-Abz/Dnp-peptide, $\lambda_{ex} = 321$ nm and $\lambda_{em} = 418$ nm.

Acknowledgments

We would like to thank Dr. Larry Stanker of the US Department of Agriculture for his monoclonal mouse antibodies and Dr. Bruce Kaplan of City of Hope for the synthesis of fluorogenic peptide substrates. This work was supported by the National Institutes of Health Grants U54 AI065359 (Pacific Southwest Regional Center of Excellence).

References

1. Gill DM (1982) Bacterial toxins: a table of lethal amounts. Microbiol Rev 46:86–94
2. Arnon SS, Schechter R, Inglesby TV, Henderson DA, Bartlett JG et al (2001) Botulinum toxin as a biological weapon: medical and public health management. JAMA 285:1059–1070
3. Long SS (2007) Infant botulism and treatment with BIG-IV (BabyBIG). Pediatr Infect Dis J 26:261–262
4. Koepke R, Sobel J, Arnon SS (2008) Global occurrence of infant botulism, 1976–2006. Pediatrics 122:e73–82
5. Werner SB, Passaro D, McGee J, Schechter R, Vugia DJ (2000) Wound botulism in California, 1951–1998: recent epidemic in heroin injectors. Clin Infect Dis 31:1018–1024
6. Sobel J, Tucker N, Sulka A, McLaughlin J, Maslanka S (2004) Foodborne botulism in the United States, 1990–2000. Emerg Infect Dis 10:1606–1611
7. Simpson LL (1981) The origin, structure, and pharmacological activity of botulinum toxin. Pharmacol Rev 33:155–188
8. Sakaguchi G (1982) *Clostridium botulinum* toxins. Pharmacol Ther 19:165–194
9. Volknandt W (1995) The synaptic vesicle and its targets. Neuroscience 64:277–300
10. Lalli G, Bohnert S, Deinhardt K, Verastegui C, Schiavo G (2003) The journey of tetanus and botulinum neurotoxins in neurons. Trends Microbiol 11:431–437
11. Schiavo G, Santucci A, Dasgupta BR, Mehta PP, Jontes J et al (1993) Botulinum neurotoxins serotypes A and E cleave SNAP-25 at distinct COOH-terminal peptide bonds. FEBS Lett 335:99–103
12. Schiavo G, Rossetto O, Catsicas S, Polverino de Laureto P, DasGupta BR et al (1993) Identification of the nerve terminal targets of botulinum neurotoxin serotypes A, D, and E. J Biol Chem 268:23784–23787
13. Schiavo G, Matteoli M, Montecucco C (2000) Neurotoxins affecting neuroexocytosis. Physiol Rev 80:717–766
14. Schiavo G, Benfenati F, Poulain B, Rossetto O, Polverino de Laureto P et al (1992) Tetanus and botulinum-B neurotoxins block neurotransmitter release by proteolytic cleavage of synaptobrevin. Nature 359:832–835
15. Schantz EJ, Johnson EA (1992) Properties and use of botulinum toxin and other microbial neurotoxins in medicine. Microbiol Rev 56:80–99
16. Johnson EA (1999) Clostridial toxins as therapeutic agents: benefits of nature's most toxic proteins. Annu Rev Microbiol 53:551–575
17. Partikian A, Mitchell WG (2007) Iatrogenic botulism in a child with spastic quadriparesis. J Child Neurol 22:1235–1237
18. Crowner BE, Brunstrom JE, Racette BA (2007) Iatrogenic botulism due to therapeutic botulinum toxin A injection in a pediatric patient. Clin Neuropharmacol 30:310–313
19. Centers for Disease Control and Prevention: Botulism in the United States, 1899–1996 (1998). Handbook for Epidemiologists, Clinicians, and Laboratory Workers. Atlanta, GA: U.S. Department of Health and Human Services, Public Health Service, CDC and Prevention. pp. 1–42
20. Schantz E, Kautter DA (1978) Standardized assay for *Clostridium botulinum* toxins. J AOAC Int 61:96–99
21. Bagramyan K, Barash JR, Arnon SS, Kalkum M (2008) Attomolar detection of botulinum toxin type A in complex biological matrices. PLoS One 3:e2041
22. Brossier F, Mock M (2001) Toxins of *Bacillus anthracis*. Toxicon 39:1747–1755
23. Hanna PC, Acosta D, Collier RJ (1993) On the role of macrophages in anthrax. Proc Natl Acad Sci U S A 90:10198–10201
24. Boyer AE, Quinn CP, Woolfitt AR, Pirkle JL, McWilliams LG et al (2007) Detection and quantification of anthrax lethal factor in serum by mass spectrometry. Anal Chem 79:8463–8470
25. Schmidt JJ, Stafford RG (2003) Fluorigenic substrates for the protease activities of botulinum neurotoxins, serotypes A, B, and F. Appl Environ Microbiol 69:297–303
26. Förster T (1948) Zwischenmolekulare Energiewanderung und Fluoreszenz. Ann Phys 437:55–75

Chapter 4

Examination of *Bacillus anthracis* Spores by Multiparameter Flow Cytometry

William C. Schumacher, Craig A. Storozuk, Prabir K. Dutta, and Andrew J. Phipps

Abstract

The ability to rapidly differentiate *Bacillus anthracis* spores from spores belonging to other *Bacillus* spp. is potentially useful for combating the intentional release of this biothreat agent. Furthermore, not all *B. anthracis* strains are fully virulent and the ability to determine the potential virulence of the endospore is also important. In this chapter, we describe a two-color flow cytometric assay capable of simultaneously identifying *B. anthracis* spores and the presence of spore-associated protective antigen, a virulence marker for strains harboring the pXO1 plasmid.

Key words: *Bacillus anthracis* spores, Flow cytometry, Protective antigen, Antibody conjugation, Fluorescence, Identification

1. Introduction

Bacillus anthracis, a Gram-positive spore-forming bacterium, is a member of the *Bacillus cereus* group (along with *B. cereus*, *B. thuringiensis*, and *B. mycoides*), which can exist ubiquitously in nature and is genetically related. One of the only distinguishing features among this group is plasmids that encode for virulence factors. Strains of *B. anthracis* can harbor two major virulence plasmids, pXO1 and pXO2 (1). The *B. anthracis* pXO1 plasmid encodes for the three proteins protective antigen (PA), edema factor (EF), and lethal factor (LF), which interact synergistically to form edema toxin (PA and EF) and lethal toxin (PA and LF). Fully virulent isolates of *B. anthracis* also harbor the pXO2 plasmid that encodes for an antiphagocytic poly-D-glutamic acid capsule. The *B. anthracis* Ames strain possesses both these virulence

Otto Holst (ed.), *Microbial Toxins: Methods and Protocols*, Methods in Molecular Biology, vol. 739, DOI 10.1007/978-1-61779-102-4_4, © Springer Science+Business Media, LLC 2011

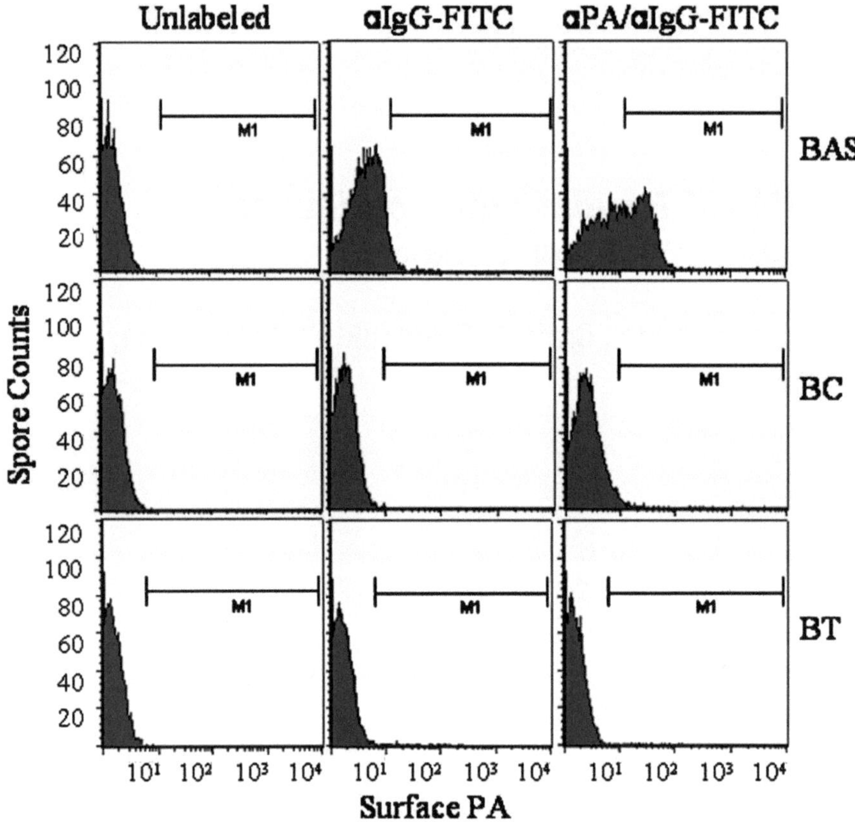

Fig. 1. Typical flow cytometry data for the surface PA assay. The monoclonal antibody against PA was found to bind tightly with *B. anthracis* Sterne spores (BAS) as compared to *B. cereus* (BC) and *B. thuringiensis* (BT) spores. Compared to the intrinsic fluorescence of unlabeled spores, IgG–FITC did nonspecifically bind to BAS. The FITC cutoff marker (M1; M1 = positive event) was adjusted to negate its contribution to the overall surface PA signal. α, anti.

plasmids, whereas minimally virulent Sterne and Pasteur strains lack pXO2 or pXO1, respectively.

Flow cytometric methods for examining *Bacillus* spp. have increasingly appeared in the literature because they work rapidly and can detect both physical and chemical features within a spore population (2–4). The technique is based on the interaction of light with spores contained in a hydrodynamically focused liquid stream. Light scattered off the spores is indicative of size and granularity, while emitted light, intrinsic or extrinsic, can reveal information about structure and function. Typical analysis speeds exceed 1,000 spores per second (5), making flow cytometry faster than the 1–2 days required for most standard microbiological tests (6). And unlike polymerase chain reaction (PCR) methods, which require the extraction of endosporal DNA (7), flow cytometric methods can rely on simple labeling assays that detect surface antigens on intact spores. However, until recently most spore-based labeling assays were not selective among members of the *B. cereus* group (8, 9).

In 2003, short peptide fragments that exhibited tight binding to *Bacillus* spores were discovered (10), and numerous studies have since reported on their effectiveness (11, 12). One such peptide, ATYPLPIRGGGC (ATYP), was conjugated to R-phycoerythrin (RPE) and found by flow cytometry to bind species specifically to *B. anthracis* spores. Unfortunately, the ATYP–RPE conjugate could not differentiate spores from different strains of *B. anthracis* (10).

Recently, the PA protein expressed by pXO1-harboring strains of *B. anthracis* was implicated as a spore surface antigen (13). PA is present during the sporulation process and thought to be non-covalently entrapped in the spore coat (SC) and exosporium (EX). Since the pXO1 plasmid encodes for several virulence factors, surface PA can be considered a virulence marker for *B. anthracis* spores. The protocol that follows describes a two-color flow cytometry assay which couples a fluorescein isothiocyanate (FITC)-conjugated antibody-based PA assay (see Fig. 1) to the *B. anthracis* spore-specific ATYP–RPE assay. We demonstrate the potential of this multiparameter assay by differentiating PA-positive (+PA) *B. anthracis* spores from PA-negative (–PA) *B. cereus* and *B. thuringiensis* spores.

2. Materials

2.1. Bacterial Stocks

1. *B. anthracis* Sterne 34F2 ($pXO1^+/pXO2^-$, seed from the live-spore veterinary vaccine; Colorado Serum Company, Denver, CO), *B. anthracis* Ames ($pXO1^+/pXO2^+$), *B. cereus* ($pXO1^-/pXO2^-$), and *B. thuringiensis* ($pXO1^-/pXO2^-$) seed lots.
2. Leighton Doi medium (LDM): 15 g bactopeptone, 6.9 g glucose, 3 g yeast extract, 3 g NaCl, 1.88 g KCl, 0.294 g $CaCl_2 \cdot 2H_2O$, 0.0246 g $MgSO_4 \cdot 6H_2O$, 0.0028 g $FeSO_4 \cdot 6H_2O$, and 0.00169 g $MnCl_2$. Adjust volume to 1 L with double-distilled water (ddH_2O) and autoclave.
3. Modified G medium (MGM): 2 g yeast extract, 2 g $(NH_4)_2SO_4$, 0.5 g K_2HPO_4, 0.2 g $MgSO_4 \cdot 6H_2O$, 0.05 g $MnSO_4 \cdot 4H_2O$, 0.025 g $CaCl_2 \cdot 2H_2O$, 0.005 g $ZnSO_4 \cdot 2H_2O$, 0.005 g $CuSO_4 \cdot 5H_2O$, and 0.0005 g $FeSO_4 \cdot 6H_2O$. Adjust volume to 1 L with ddH_2O and autoclave.
4. Renografin 76 sedimentation gradient (see Subheading 3.2, items 2 and 3).
5. Sterile H_2O, Gram's stain, blood agar plates, sterile loops, sterile laboratory glassware, Pasteur pipettes, and sterile centrifuge tubes.
6. Equipment: autoclave, box shaker, incubation chamber, centrifuge, hemocytometer, and optical microscope.

2.2. Two-Color Assay

1. Mouse monoclonal antibody against *B. anthracis* PA (anti-PA, 2 mg/mL, Abcam, Cambridge, MA).
2. FITC-conjugated goat anti-mouse immunoglobulin G (IgG), H+L (IgG–FITC, Jackson ImmunoResearch, West Grove, PA).
3. Synthetic peptide with amino acid sequence ATYPLPIRG GGC (GenScript, Piscataway, NJ).
4. RPE (4 mg/mL, Invitrogen, Carlsbad, CA).
5. Sulfo-succinimidyl-4-(*N*-maleimidomethyl)cyclohexane-1-carboxylate (Sulfo-SMCC, Pierce, Rockford, IL).
6. Staining buffer: Dulbecco's calcium-/magnesium-free phosphate-buffered saline (PBS), pH 7.2, 1% fetal bovine serum.
7. Wash buffer: Dulbecco's PBS, pH 7.2.
8. 2% Paraformaldehyde in wash buffer.
9. Microcentrifuge tubes, Slide-A-Lyzer 10K MWCO dialysis cassettes, and sterile 96-well plates (conical bottom).
10. Equipment: Vortex mixer, micropipettors, microcentrifuge, incubation chamber, flow cytometer, and laser-scanning confocal microscope.

3. Methods

3.1. Preparation of Bacterial Stocks

1. Streak blood agar plate with a sample of the bacterial seed lot using a sterile loop and incubate for 12–18 h at 37°C.
2. Using a sterile loop, remove a single colony off the cultured plate and add to 15–20 mL LDM; incubate with shaking for 12–18 h at 37°C.
3. Transfer the 15–20 mL culture into a 200 mL volume of LDM and incubate with shaking for 10 h at 37°C.
4. Centrifuge cultured bacteria for 20 min at 4,000 × *g* and 4°C, then discard supernatant and transfer pellet into 250 mL MGM; cover container with aluminum foil and shake gently for 3–5 days at 20–25°C.
5. Mix an aliquot of the MGM (containing spores) with Gram's stain in a 1:1 ratio (v/v) and evaluate the extent of sporulation with a hemocytometer or a Petroff-Houser counter. If sporulation is complete and there is less than 10–15% vegetative cells, then proceed to purification; if there is greater than 15% vegetative cells, continue the incubation at room temperature for an additional 24 h.

3.2. Purification of Bacterial Stocks

1. Centrifuge the MGM (containing spores) for 20 min at 4,000 × *g* and 4°C, discarding the supernatant.

2. Prepare a stock solution of Renografin 76 by mixing 16 mL of filter sterilized meglumine diatrozoate (833 mg/mL) with 2 mL of sodium diatrozoate in ddH_2O. Adjust the density of the Renografin 76 stock solution to 1.25 g/mL using additional ddH_2O.
3. Divide the stock Renografin 76 solution into two equal volumes (stock A and B); stock A = bottom phase of gradient (do not modify). Adjust the density of stock B to 1.12 g/mL using additional ddH_2O.
4. Pipette 10 mL stock B into sterile centrifuge tube containing the crude spore pellet, then vortex. Transfer the entire volume into a sterile Oak Ridge centrifuge tube, and then carefully pipette 10 mL of stock A to the bottom of the centrifuge tube, being sure not to disturb the density interface.
5. Centrifuge for 1 h at 8,000 × *g*. The pellet contains mature *B. anthracis* spores. Remove the Renografin density gradient and any bands which contain immature spores and vegetative cells, leaving the pellet. Wash the pellet three times with ddH_2O, then reconstitute in 3 mL sterile H_2O, and store at 4°C.
6. Using the serial dilution method, determine the colony forming units per milliliter (CFU/mL) of the purified spore preparation.

3.3. Preparation of ATYP–RPE Conjugate

1. Centrifuge 50 μL RPE for 5 min at 9,000 × *g* (at 20–25°C), then discard the supernatant, and reconstitute the pellet in 25 μL wash buffer.
2. Dialyze RPE against several changes of wash buffer to remove residual ammonium sulfate (total volume after 4 h of dialysis at 20–25°C ≈ 30 μL).
3. Dissolve 2 mg of sulfo-SMCC into 450 μL wash buffer (10.2 mM sulfo-SMCC).
4. Mix 30 μL RPE with 15 μL sulfo-SMCC stock solution (molar ratio ≈ 1 RPE: 180 sulfo-SMCC) and incubate for 1 h at 30°C (see Note 1). Remove unbound sulfo-SMCC via dialysis into wash buffer (total volume after 6 h of dialysis at 4°C ≈ 50 μL).
5. Dissolve 2 mg of ATYP peptide into 200 μL wash buffer (8.3 mM ATYP peptide).
6. Mix 50 μL sulfo-SMCC-activated RPE with 100 μL ATYP peptide stock solution (molar ratio ≈ 1 RPE: 1,000 ATYP peptide) and incubate for 18 h at 30°C. Remove unbound ATYP peptide via dialysis into wash buffer (total volume after 18 h of dialysis at 4°C ≈ 200 μL). This corresponds to approximately 4 μM of purified ATYP–RPE conjugate.

3.4. Two-Color Assay

1. Pipette a volume containing approximately 10^6 spores into a sterile 96-well plate, add staining buffer and centrifuge at 700×*g* and 4°C, discarding the supernatant. To the spore pellet, add 25 μL of a 1:12.5 dilution of the anti-PA antibody (in staining buffer), then incubate for 1 h at 37°C. Remove unbound anti-PA antibody via repeated wash/centrifuge steps (at 700×*g* and 4°C) with 200 μL aliquots of wash buffer.
2. Mix spores with 25 μL of a 1:100 dilution (in wash buffer) of IgG–FITC for 1 h at 37°C. Remove unbound anti-IgG–FITC via repeated wash/centrifuge steps (at 700×*g* and 4°C) with 200 μL aliquots of wash buffer.
3. Mix spores with 25 μL of 2 μM ATYP–RPE and incubate for 1 h at 37°C. Remove unbound conjugate via repeated wash/centrifuge steps (at 700×*g* and 4°C) with wash buffer (see Note 2).
4. Fix spores in 2% PFA prior to analysis by flow cytometry (see Note 3) and laser-scanning confocal microscopy (see Note 4).

Confocal microscopy was used to visually confirm that double-labeling occurred on Sterne spores and not single-color labeling of different populations (see Fig. 2). We next examined several spore lots by flow cytometry to see if differences in preparation, purification, and storage conditions affected the outcome of the assay (see Table 1). Lots BAS 3 through BAS8 were labeled similarly by ATYP–RPE (51–71% positive) and had comparable surface PA values (23–43% positive). Only lots BAS1 and BAS2 demonstrated lower ATYP–RPE binding (11–12% positive) and higher surface PA values (72–75% positive). Lots BAS1 and BAS2 were stored in ddH_2O at 4°C for a longer period of time (4 years) than any other lot used in this study (see Note 5 and Fig. 3). As expected, *B. cereus* and *B. thuringiensis* spores produced double-negative results by the two-color assay.

Based on the two-color results, we developed a set of prototypical flow cytometry dot-plot patterns for predicting the species of spores belonging to the *Bacillus* genus and their potential virulence (see Fig. 4). Pattern 1 (events occur in lower-left quadrant) corresponds to ATYP–RPE$^-$/PA$^-$ non-*B. anthracis* spores, such as *B. cereus* or *B. thuringiensis.* Pattern 2 (events occur in lower-left, upper-left, and upper-right quadrants) is indicative of ATYP–RPE$^+$/PA$^+$ spores, such as *B. anthracis* Sterne or Ames, that are potentially virulent. Pattern 3 (events occur in lower-left and upper-left quadrants) is indicative of ATYP–RPE$^+$/PA$^-$ spores, such as *B. anthracis* Pasteur, that are likely to be minimally virulent. Like pattern 2, pattern 4 (events occur in all four quadrants) also results from ATYP–RPE$^+$/PA$^+$ spores that are potentially

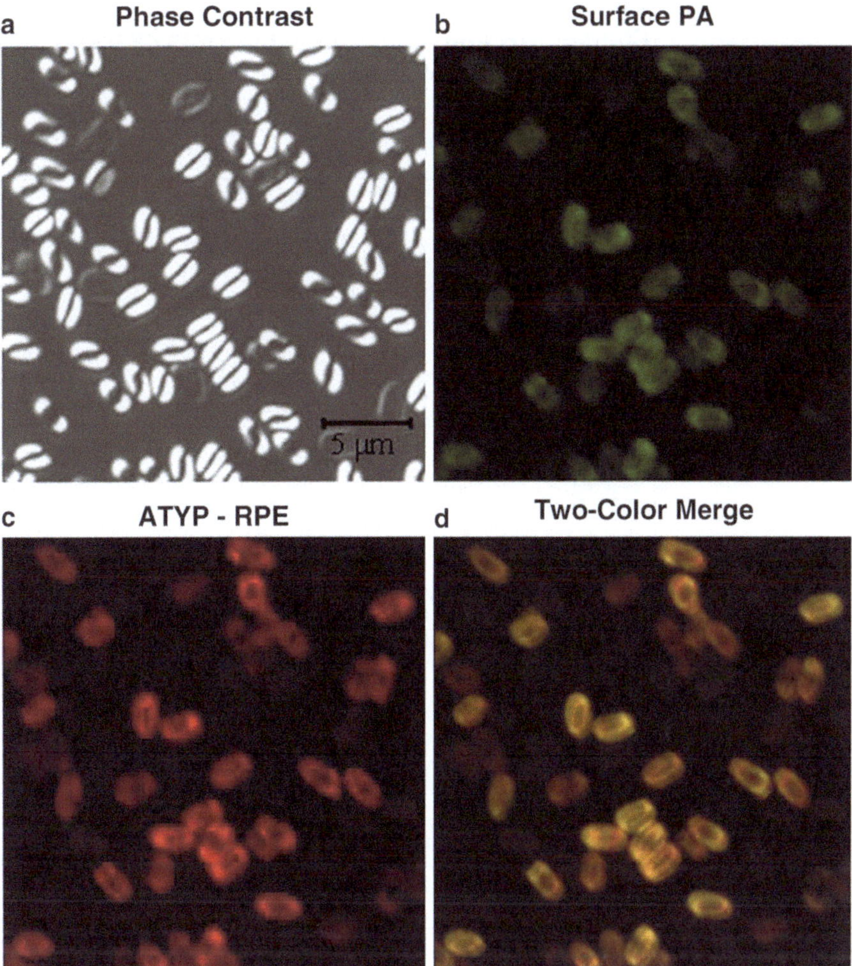

Fig. 2. Phase contrast (**a**) and fluorescence (**b** and **c**) images of double-labeled *B. anthracis* Sterne spores. The florescence observed in image (**b**) was attributed to IgG–FITC binding with anti-PA/surface PA complexes. The florescence in image (**c**) was caused by surface-bound ATYP–RPE. Fluorescence images (**b** and **c**) were used to produce the merged image (**d**), which confirmed that double-labeling occurred on most Sterne spores.

virulent; however, the large population in the lower right quadrant indicates that a significant fraction of the spores containing PA are nonreactive with the ATYP–RPE conjugate.

In conclusion, this two-color assay successfully differentiated $pXO1^+$ (+PA) strains of *B. anthracis* from $pXO1^-$ (–PA) strains in only a few hours. One drawback of this assay is that it cannot resolve *B. anthracis* Sterne spores (BAS) from *B. anthracis* Ames spores, since they both harbor the pXO1 plasmid. Nonetheless, this two-parameter detection assay marks an important step toward rapid and complete characterization of this dangerous pathogen.

Table 1
Characterization of different spore lots by using the flow cytometry two-color assay

Organism	Lot	D.O.P.[a]	%ATYP–RPE	%PA+	%DP[b]
B. cereus	BC1[c]	04/2007	0.070	1.5	0.03
B. thuringiensis	BT1[c]	04/2007	0.26	0.16	0.08
B. anthracis Sterne	BAS1[d]	09/2003	12	72	8.6
	BAS2[d]	09/2003	11	75	6.8
	BAS3[d]	06/2005	58	36	36
	BAS4[c]	08/2005	58	33	33
	BAS5[c]	06/2006	51	23	23
	BAS6[e]	05/2007	67	32	32
	BAS7[c]	05/2007	56	30	30
	BAS8[d]	06/2007	71	43	43

[a] D.O.P. = date (month/year) of spore lot preparation
[b] %DP = percentage of spores that are double positive
[c] Leighton Doi medium/modified G medium/Renografin 76
[d] Leighton Doi medium/nutrient agar plate/Renografin 76
[e] Leighton Doi medium/modified G medium/no Renografin 76

4. Notes

1. Sulfo-SMCC (MW = 436.37 g/mol) is a heterobifunctional cross-linker containing an NHS ester group (amine-reactive at pH 7–9) and a maleimide group (sulfhydryl-reactive at pH 6.5–7.5). Sulfo-SMCC is soluble in water at concentrations up to ≈10 mM. Sulfo-SMCC-activated materials should be used immediately after purification to minimize maleimide decomposition caused by neutral pH.
2. Lower surface PA values were obtained for Sterne spores when ATYP–RPE was incubated before, or simultaneous to, IgG–FITC (data not shown). This finding may indicate that surface PA exists close to the ATYP-peptide receptor(s) on the exosporium basal layer and can be blocked by binding of ATYP–RPE.
3. Flow cytometry: Samples were analyzed using an FACSCalibur instrument and CellQuest Pro software (Becton Dickinson Biosciences). All samples were excited at 488 nm with an

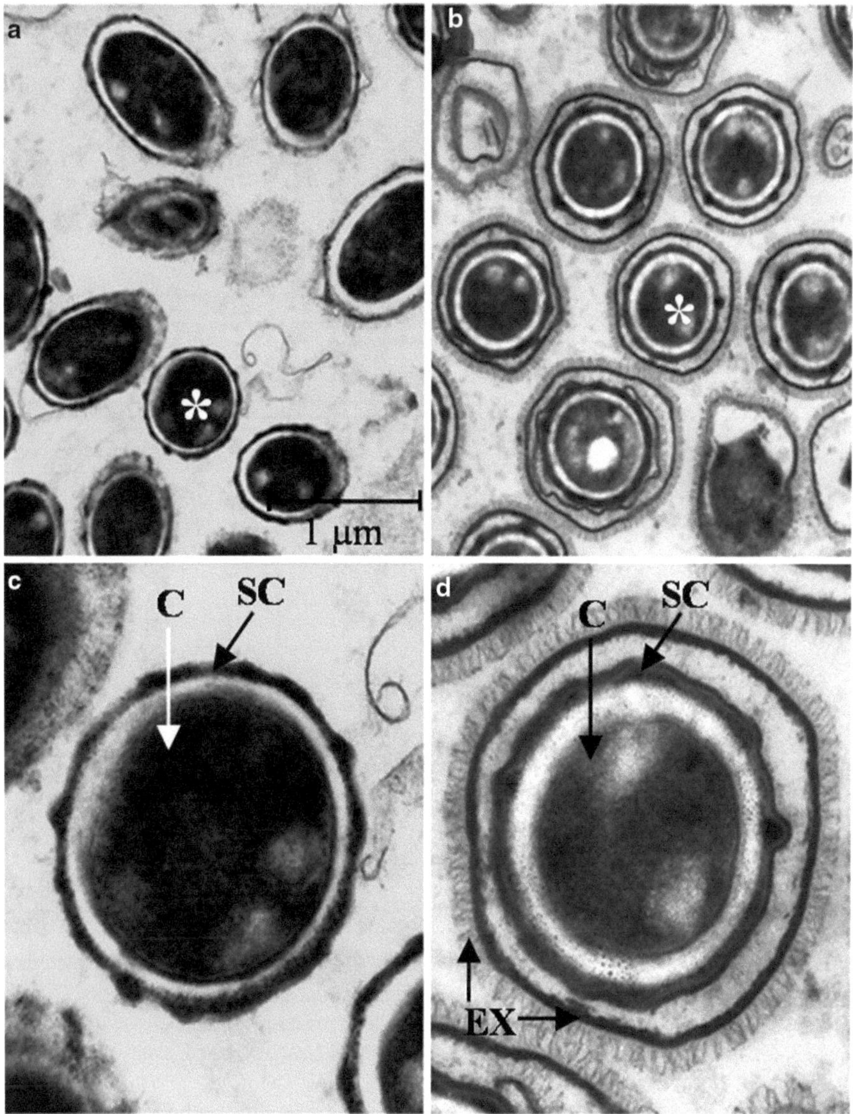

Fig. 3. Transmission electron micrographs (see Note 6) of spores from lots BAS1 (**a** and **c**) and BAS8 (**b** and **d**). Asterisks are placed on spores shown in the enlarged images. Arrowheads point to the exosporium (EX), spore coat (SC), and core (C). Image magnifications are ×21,300 (**a** and **b**) and ×60,000 (**c** and **d**).

argon-ion laser and detected through FL1 (FITC; 530 ± 15 nm) and FL2 (PE; 585 ± 21 nm) bandpass filters. Unlabeled spores and negative controls were used to empirically determine voltages, gains, and cutoff values for positive samples. Representative instrument settings were as follows: FL1 voltage/gain = 545/1.00, FL2 voltage/gain = 543/1.00, FITC cutoff marker = <2.0%, and RPE cutoff marker = <0.50%. Typical compensation values for two-color assays were

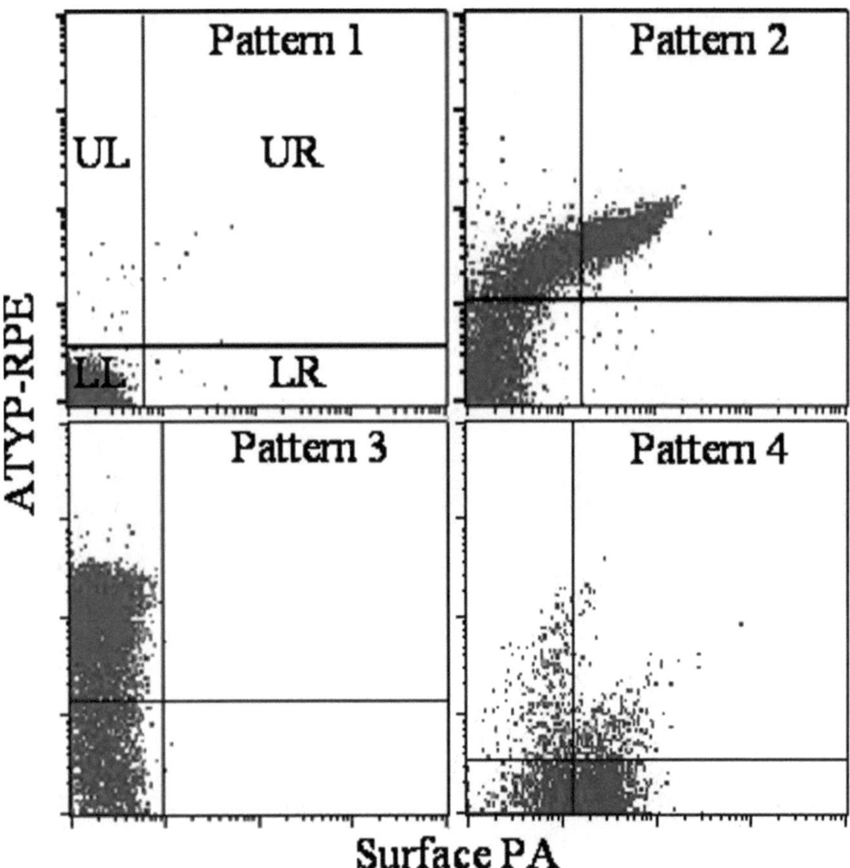

Fig. 4. Prototypical flow cytometry dot-plot patterns for identification and characterization of samples suspected to contain *Bacillus* spores. Species and potential virulence can be predicted by using one of the following patterns: pattern 1, ATYP–RPE$^-$/PA$^-$; pattern 2, ATYP–RPE$^+$/PA$^+$; pattern 3, ATYP–RPE$^+$/PA$^-$; and pattern 4, ATYP–RPE$^+$/PA$^+$ (partially nonreactive with ATYP–RPE). Quadrant abbreviations are as follows: *LL* lower left, *UL* upper left, *UR* upper right, and *LR* lower right. The following *Bacillus* samples were used to generate the dot-plot patterns: pattern 1, *B. thuringiensis*; pattern 2, *B. anthracis* Sterne; pattern 3, *B. anthracis* Sterne labeled with ATYP–RPE only (the *B. anthracis* Pasteur strain was not available for this study); and pattern 4, *B. anthracis* Sterne, lot BAS1 (no exosporium).

FL1 – 1.9% FL2, and FL2 – 17.6% FL1. All flow cytometry data was presented as an average of 10,000 events.

4. Confocal microscopy: Samples were analyzed using a Leica TCS SP2 AOBS confocal laser scanning microscope. An aliquot (10 μL) of spore suspension was deposited on a clean glass microscope slide and dried to immobilize the spores. Dried samples were mounted with 90% glycerol, then coverslipped and imaged immediately. All samples were excited at 488 nm with an argon-ion laser and detected using FL1 (FITC; 520 ± 10 nm) and FL2 (PE; 575 ± 10 nm). Unlabeled spores and negative controls were used to empirically determine voltages for positive samples, which were typically 450 and 600 for FL1 and FL2, respectively. Using these optimized

settings, we did not observe any spectral bleed-through into either PMT. All images were collected using a 40× oil objective lens and presented as an average of ten scans.

5. Transmission electron microscopic (TEM) examination of spores from lots BAS1 and BAS8 revealed that a majority of lot BAS1 spores were missing the outermost exosporium. We believe the structural damage to lot BAS1 spores was caused by either repeated handling during the prolonged storage or unknown protease activity (possibly through contamination).
6. TEM: Unlabeled spores were prefixed with 5% gluteraldehyde in 20 mM sodium phosphate buffer, pH 7.2 for 3 days at 20–25°C. Postfixation was carried out with 1% osmium tetroxide in 50 mM sodium phosphate buffer, pH 7.2 for 24 h at 20–25°C. Serial alcohol dehydration was then performed using ethanol (15, 35, 50, 75, 90, and 100%, 3×) and then acetone (100%, 3×). Next, the samples were embedded with epon resin and polymerized at 75°C for 24 h. Ultrathin sections were obtained using an LKB Ultramicrotome (Sweden). The sections were double-stained with 1% uranyl acetate and lead citrate. EM observation was made by a Phillips EM 300 transmission electron microscope at an accelerating voltage of 60 kV. The images were recorded on Kodak electron image films.

Acknowledgments

This material is based upon the work supported by the National Science Foundation under grant no. 0221678. We are grateful to Mamoru Yamaguchi for performing the electron microscopy work. The *B. anthracis* Ames, *B. cereus*, and *B. thuringiensis* seed lots were generous gifts from Battelle, Columbus, OH.

References

1. Edwards KA, Clancy HA, Baeumner AJ (2006) *Bacillus anthracis*: toxicology, epidemiology and current rapid-detection methods. Anal Bioanal Chem 384:73–84
2. Leser TD, Knarreborg A, Worm J (2008) Germination and outgrowth of *Bacillus subtilis* and *Bacillus licheniformis* spores in the gastrointestinal tract of pigs. J Appl Microbiol 104:1025–1033
3. Kim J-H, Roh C, Lee C-W, Kyung D, Choi S-K, Jung H-C, Pan J-G, Kim B-G (2007) Bacterial surface display of GPF_{uv} on *Bacillus subtilis* spores. J Microbiol Biotechnol 17:677–680
4. Cronin UP, Wilkinson MG (2009) The potential of flow cytometry in the study of *Bacillus cereus*. J Appl Microbiol 108(1):1–16
5. Nebe-von-Caron G, Stephens PJ, Hewitt CJ, Powell JR, Badley RA (2000) Analysis of bacterial function by multi-colour fluorescence flow cytometry and single cell sorting. J Microbiol Methods 42:97–114
6. Marston CK, Gee JE, Popovic T, Hoffmaster AR (2006) Molecular approaches to identify

and differentiate *Bacillus anthracis* from phenotypically similar *Bacillus* species isolates. BMC Microbiol 6:1–7

7. Ramisse V, Patra G, Garrigue H, Guesdon J-L, Mock M (1996) Identification and characterization of Bacillus anthracis by multiplex PCR analysis of sequences on plasmids pXO1 and pXO2 and chromosomal DNA. FEMS Microbiol Lett 145:9–16
8. Kamboj DV, Agarwal GS, Dwarkanath BS, Adhikari SI, Singh L (2006) Flow-cytometric analysis of *Bacillus anthracis* spores. Def Sci J 56:769–774
9. Stopa PJ (2000) The flow cytometry of *Bacillus anthracis* spores revisited. Cytometry 41:237–244
10. Williams DD, Benedek O, Turnbough CL Jr (2003) Species-specific peptide ligands for the detection of *Bacillus anthracis* spores. Appl Environ Microbiol 69:6288–6293
11. Acharya G, Doorneweerd DD, Chang C, Henne WA, Low Pl S, Savran CA (2007) Label-free optical detection of anthrax-causing spores. J Am Chem Soc 129: 732–733
12. Pai S, Ellington AD, Levy M (2005) Proximity ligation assays with peptide conjugate "burrs" for the sensitive detection of spores. Nucleic Acids Res 33:1–7
13. Cote CK, Rossi CA, Kang AS, Morrow PR, Lee JS, Welkos SL (2005) The detection of protective antigen (PA) associated with spores of *Bacillus anthracis* and the effects of anti-PA antibodies on spore germination and macrophage interactions. Microb Pathog 38:209–225

Chapter 5

A Cell-Based Fluorescent Assay to Detect the Activity of Shiga Toxin and Other Toxins that Inhibit Protein Synthesis

Shane Massey, Beatriz Quiñones, and Ken Teter

Abstract

Escherichia coli O157:H7, a major cause of food-borne illness, produces Shiga toxins (Stxs) that block protein synthesis by inactivating the ribosome. In this chapter, we describe a simple cell-based fluorescent assay to detect Stxs and inhibitors of toxin activity. The assay can also be used to detect other plant and bacterial toxins that arrest protein synthesis.

Key words: Food-borne pathogen, Food safety, Shiga toxin, Toxin detection, Toxin inhibitors, Toxicity assay, Vero cells

1. Introduction

Shiga toxins (Stxs) are produced by enterohemorrhagic *Escherichia coli*, a food- and water-borne pathogen that in the USA alone causes an estimated 73,000 cases of illness per year (1). These toxins remove a single adenine residue from the 28S rRNA of the eukaryotic 60S ribosomal subunit (2). The resulting inhibition of protein synthesis generates a cytotoxic effect.

Several methods can be used to detect Stx activity against mammalian cells, but there are disadvantages to each technique. A common procedure measures the viability of intoxicated cells by dye exclusion, MTT assay, or similar protocols (3–6). This approach requires several days of incubation and often produces poor quantitative data. Another more quantitative method uses the incorporation of radiolabeled amino acids into newly synthesized proteins to measure the Stx-induced inhibition of protein synthesis (7, 8). This assay requires the handling of radioactive

Otto Holst (ed.), *Microbial Toxins: Methods and Protocols*, Methods in Molecular Biology, vol. 739,
DOI 10.1007/978-1-61779-102-4_5,

material, which is laborious, and can only process a limited number of samples. A quantitative luciferase-based assay has been described which is similar to the system reported here, but the luciferase assay requires several preparatory and processing steps to enact the detection method (9). Nonradioactive cell-free assays have also been developed, yet these assays require high concentrations of toxin and/or commercially purchased kits to detect the Stx-induced inhibition of protein synthesis (10, 11).

As an alternative to existing technologies, we developed a cell-based assay for the detection of Stx. A Vero cell line with constitutive expression of a destabilized variant ($t_{1/2}$ = 2 h) of the enhanced green fluorescent protein (d2EGFP) is incubated 16–18 h with purified toxin or cell-free culture supernatants from Stx-producing bacteria (see Note 1). Toxin-treated cells degrade d2EGFP and do not produce more of the protein due to the toxin-induced block of protein synthesis. The fluorescent signal from Vero-d2EGFP cells is accordingly lost in proportion to the applied dose of toxin. This simple, quantitative assay provides reproducible data with minimal hands-on effort (see Note 2). The procedure does not require radioisotopes, commercial kits, or additional processing steps. A fluorescent plate reader is required for sample reading, but the only major recurring cost is the use of black 96-well microplates with clear bottoms. As described below, the Vero-d2EGFP assay can also be adapted to screen for toxin inhibitors or to detect other toxins that inhibit protein synthesis.

2. Materials

2.1. Mammalian Cell Culture

1. Complete medium: Dulbecco's Modified Eagle Medium (DMEM) high glucose with L-glutamine and sodium pyruvate (GIBCO, Grand Island, NY) supplemented with 10% fetal bovine serum (FBS) and 1% antibiotic–antimycotic solution (GIBCO).
2. Antibiotic–antimycotic 100× solution containing 100 U/mL penicillin G sodium, 100 μg/mL streptomycin sulfate, 25 μg/mL amphotericin B as Fungizone® in 0.85% saline (GIBCO).
3. Freeze medium: 50% DMEM, 40% FBS, and 10% DMSO.
4. F-12 + GlutaMAX-1 nutrient mixture (Ham's F-12; GIBCO).
5. 0.05% Trypsin–EDTA 1× (GIBCO).
6. Phosphate buffered saline (PBS), 10× stock consisting of 0.58 M Na_2HPO_4, 0.17 M NaH_2PO_4, and 0.69 M NaCl.

Prepare 1× working stock by dilution of one part with nine parts water and sterilize by autoclaving.

7. Greiner Bio-One CELLSTAR® black 96-well polystyrene plates with a µClear® flat bottom.
8. Geneticin (Gibco).
9. 100×20 mm tissue culture dishes.
10. Grenier Bio-One CELLSTAR® clear 24-well and 6-well polystyrene plates.
11. Cloning cylinders (Fisher Scientific, Fair Lawn, NJ).
12. Fisherbrand 50 mL sterile pipette basins (Fisher Scientific).
13. Multichannel p200 pipette (Rainin, Oakland, CA).

2.2. Toxins

1. Stx2 (List Biological Laboratories, Campbell, CA, and Sigma–Aldrich, St. Louis, MO) (see Note 3).
2. Diphtheria toxin (DT) (List Biological Laboratories).
3. Ricin (Vector Laboratories, Burlingame, CA).
4. Exotoxin A (List Biological Laboratories).

3. Methods

We generated a Vero cell line that stably expresses d2EGFP-N1, an EGFP variant that contains a C-terminal PEST sequence for rapid proteasomal degradation (see Note 4). Stxs inhibit protein synthesis, so toxin-susceptible cells degrade d2EGFP and do not produce more of the protein. Productive intoxication accordingly results in a loss of fluorescence from the Vero-d2EGFP cells. As shown in Fig. 1, Stx2 attenuated the Vero-d2EGFP fluorescent signal in a dose-dependent manner. Furthermore, the Stx-induced loss of EGFP fluorescence mirrored the Stx-induced inhibition of overall protein synthesis as assessed by the incorporation of [^{35}S] methionine into newly synthesized proteins. Our results also demonstrated that the Vero-d2EGFP assay can readily detect picogram per milliliter concentrations of Stx.

Disruptions to the intoxication process permit the continued synthesis of d2EGFP in toxin-treated cells. This principle was demonstrated in a pilot screen to identify toxin inhibitors from a select panel of 12 plant compounds (12). A subset of these experimental results is shown in Fig. 2a. With this screen, we found that grape seed and grape pomace extracts both provided strong cellular protection against Stx. In the presence of these compounds, the Vero-d2EGFP cells withstood the effects of Stx and maintained a strong fluorescent signal through the continued synthesis of d2EGFP. The antitoxin properties of the grape

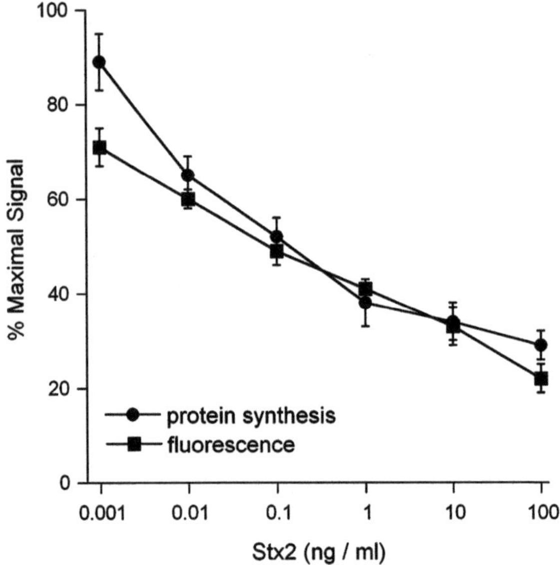

Fig. 1. Effect of Stx2 on Vero-d2EGFP fluorescence and overall protein synthesis. Protein synthesis and fluorescence were measured in separate samples of Vero-d2EGFP cells after a 16-h incubation with the stated concentrations of Stx2. The means ± standard errors of the means of at least three independent experiments with triplicate samples for each condition are shown. Reproduced from Quiñones et al. (12) with permission from the American Society for Microbiology.

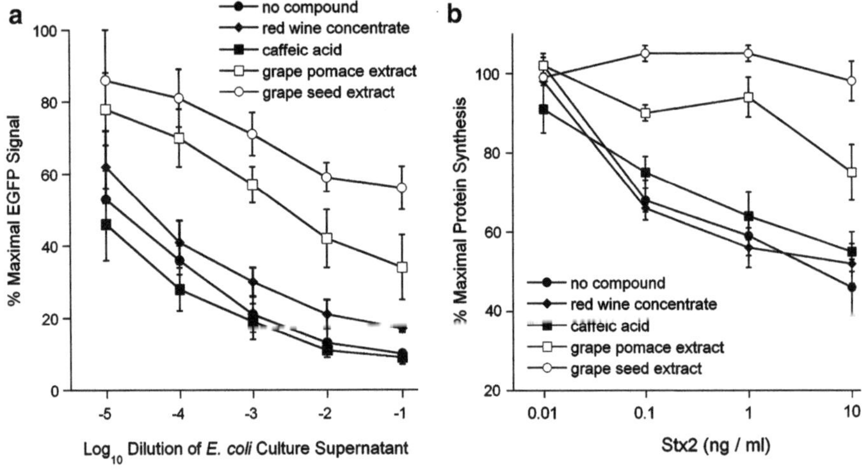

Fig. 2. Use of the Vero-d2EGFP assay to screen for toxin inhibitors. (**a**) Vero-d2EGFP cells were incubated for 16 h with plant compounds and various dilutions of a cell-free culture supernatant from an Stx-expressing strain of *E. coli* 0157:H7. Fluorescent output was then recorded with a plate reader. Cells were co-incubated with no plant compound, 0.5 mg/mL of caffeic acid, 1 mg/mL of red wine concentrate, 0.5 mg/mL of grape pomace extract, or 0.5 mg/mL of grape seed extract. The averages ± standard deviations of three independent experiments with triplicate samples for each condition are shown. (**b**) Protein synthesis was measured in Vero-d2EGFP cells after a 2-h co-incubation with plant compounds and the stated dilutions of Stx2. Cells were co-incubated with no plant compound, 1 mg/mL of caffeic acid, 1 mg/mL of red wine concentrate, 0.5 mg/mL of grape pomace extract, or 0.1 mg/mL of grape seed extract. The means ± standard errors of the means of at least four independent experiments with triplicate samples for each condition are shown. Reproduced from Quiñones et al. (12) with permission from the American Society for Microbiology.

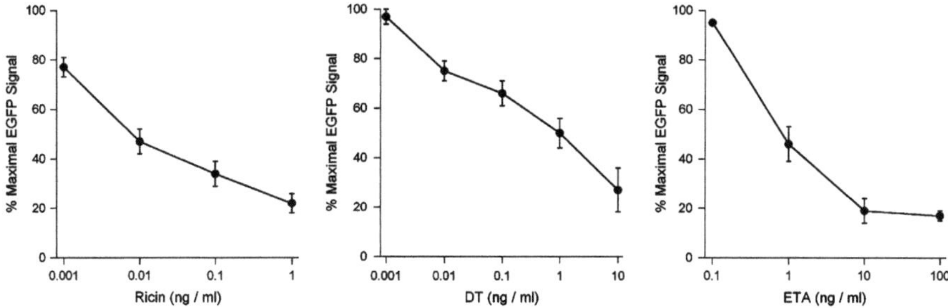

Fig. 3. Effect of ricin, diphtheria toxin, and exotoxin A on Vero-d2EGFP fluorescence. The fluorescent output from Vero-d2EGFP cells was recorded after a 16-h incubation with the stated concentrations of ricin, diphtheria toxin (DT), or exotoxin A (ETA). The means ± standard errors of the means of at least three independent experiments with triplicate samples for each condition are shown.

extracts were confirmed with an independent toxicity assay that monitored the overall level of protein synthesis in cells exposed to Stx2 (Fig. 2b) (see Note 5).

In theory, the Vero-d2EGFP assay could be used to detect the activity of any toxin that inhibits protein synthesis. We established this principle by using the Vero-d2EGFP assay to generate dose-response curves for three other toxins that inactive ribosome function: ricin, diphtheria toxin, and *Pseudomonas aeruginosa* exotoxin A (Fig. 3). Each toxin attenuated the Vero-d2EGFP fluorescent signal in a dose-dependent manner.

3.1. Transfection of Vero Cells

1. Vero CCL-81 cells (ATCC, Manassas, VA) are passaged when the cells reach a >90% confluent monolayer on a 10-cm dish.
2. Working in a tissue culture hood, the spent medium is removed from the tissue culture dish and the cells are washed twice with 10 mL of sterile 1× PBS.
3. Cells are detached from the dish by adding 1 mL of trypsin–EDTA for 5 min at 37°C.
4. Detached cells are resuspended in 9 mL of complete medium for a total volume of 10 mL.
5. 1 mL of cell suspension is added to 0.5 mL of complete medium in a well of a six-well plate.
6. Cells are grown 16–18 h to reach ~80% confluency (see Note 6).
7. The cells are transfected by mixing in a sterile 1.5-mL microcentrifuge tube 1 μg of pd2EGFP-N1 plasmid (BD Biosciences, Palo Alto, CA) with 100 μL of DMEM lacking FBS, antibiotic, and antimycotic. In a separate sterile 1.5-mL microcentrifuge tube, 5 μL of Lipofectamine reagent (Invitrogen, Carlsbad, CA) is mixed with 100 μL of DMEM lacking FBS, antibiotic, and antimycotic (see Note 7).

8. The two mixtures are combined and allowed to sit at 25°C for 30 min.
9. During the 30-min incubation for the plasmid–Lipofectamine mixture, the Vero cells are washed twice with 1 mL DMEM and then bathed in 1 mL DMEM lacking FBS, antibiotic, and antimycotic.
10. The plasmid–Lipofectamine mixture is added to the Vero cells for 4 h at 37°C.
11. The transfection medium is replaced with 1 mL of DMEM containing 10% FBS but lacking antibiotic and antimycotic.
12. The transfected cells are returned to the 37°C tissue culture incubator.

3.2. Selection of the Vero-d2EGFP Cell Line

1. At 24 h posttransfection, the cells are lifted from the six-well plate with 0.25 mL trypsin–EDTA and mixed with 0.75 mL of complete medium.
2. 0.2 mL of the cell suspension is added to a 10-cm dish containing 10 mL complete medium. Five 10 cm dishes are seeded in this manner.
3. After a 16–18 h incubation, 1 mg/mL geneticin is added to each dish.
4. The geneticin-containing medium is replaced every 3 days until colonies are visible.
5. Individual colonies are lifted with cloning cylinders using 0.1 mL trypsin–EDTA.
6. The cell suspension is placed in the well of a 24-well plate that contains 0.5 mL of complete medium. The cells are grown under selective pressure until they reach confluency.
7. Cells are lifted from the 24-well plate with 250 μL trypsin–EDTA and placed in the well of a six-well plate that contains 1 mL of complete medium. The cells are grown under selective pressure until they reach confluency.
8. Cells are lifted from the six-well plate with 500 μL trypsin–EDTA and placed in a 10-cm dish that contains 10 mL of complete medium. The cells are grown under selective pressure until they reach confluency.
9. At this point, each 10 cm dish represents a single clone from the original selection dish. When the cells in an individual dish reach confluency, they are lifted from the dish with 1 mL trypsin–EDTA and resuspended in 9 mL complete medium.
10. For each cell suspension, an FACScalibur (BD, Franklin Lakes, NJ) cell sorter is used to isolate the subpopulation of cells with the strongest fluorescent output. The 10% of cells with the highest fluorescent signal are collected and seeded to

a fresh 10-cm dish containing 10 mL of complete medium. d2EGFP expression levels are maintained by the addition of 1 mg/mL geneticin during each passage of the cell line(s).

11. Data from the FACScalibur cell sorter is used to identify a single cell line with high expression levels of d2EGFP. This Vero-d2EGFP cell line is used for all experimental studies (see Note 8).

3.3. Plating of Cells for the Fluorescence Assay

1. The Vero-d2EGFP and parental Vero CCL-81 cells are passaged when the cells reach a >90% confluent monolayer on a 10-cm dish (see Note 9).
2. Working in a tissue culture hood, the spent media is removed from the tissue culture dish and the cells are washed twice with 10 mL of sterile 1× PBS.
3. Cells are detached from the dish by adding 1 mL of trypsin–EDTA for 5 min at 37°C.
4. Detached cells are resuspended in 9 mL of complete medium for a total volume of 10 mL.
5. The cell concentration is determined with a hemocytometer and altered to a final concentration of 100,000 cells/mL. For the Vero CCL-81 cells, the final cell suspension should be in a 1.5-mL volume per 96-well plate. For the Vero-d2EGFP cells, the final cell suspension should be in a 9-mL volume per 96-well plate.
6. Each cell dilution (Vero-d2EGFP and Vero CCL-81) is poured into a separate sterile trough. Using a multichannel pipette (p200), 100 μL of the Vero CCL-81 cell suspension is transferred to the first row (12 wells) of a Greiner black 96-well microplate with clear bottom (see Note 10). Again, using a p200 multichannel pipette, 100 μL of the Vero-d2EGFP cell suspension is transferred to each of the remaining wells of the microplate.
7. Cells are grown for 24–48 h at 37°C in a 5% CO_2 humidified incubator (see Note 11).

3.4. Treatment of the Cells with Toxin Inhibitors and/or Purified Toxin

1. Several tenfold serial dilutions of toxin are prepared in serum-free Ham's F-12 medium (see Note 12). The final volume for each toxin dilution is 0.5 mL for triplicate samples, 1 mL for 6 replicate samples, or 1.5 mL for 12 replicate samples (see Note 13). For a screen of toxin inhibitors, identical serial dilutions are prepared in Ham's F-12 medium containing the final concentration of inhibitor.
2. Medium is removed from the Vero and Vero-d2EGFP cells with a p200 multichannel pipette.
3. Vero cells are incubated with 100 μL per well of serum- and toxin-free Ham's F-12 medium.

4. Vero-d2EGFP cells are incubated with 100 μL per well of toxin serial dilutions in the absence or presence of inhibitor.
5. As an unintoxicated control condition, Vero-d2EGFP cells are incubated with 100 μL per well of serum- and toxin-free Ham's F-12 medium.
6. As a positive control for the loss of fluorescence, one set of unintoxicated Vero-d2EGFP cells can be treated with 100 μg/mL of the protein synthesis inhibitor cycloheximide (see Note 14).
7. Cells are incubated with toxin for 16–18 h before fluorescence measurements are taken.

3.5. Fluorescence Measurements and Data Analysis

1. Using a p200 multichannel pipette, remove the 100 μL of spent medium from each well.
2. Using a p200 multichannel pipette, wash the cells twice with 100 μL of 1× PBS (see Note 15).
3. Using a p200 multichannel pipette, add another 100 μL of 1× PBS to the cells. Measurements are read from cells bathed in 1× PBS (see Note 16).
4. EGFP fluorescence is measured on a Synergy HT Multi-Detection Microplate Reader (BioTek, Winooski, VT) with the 485/20 nm excitation filter and the 528/20 nm emission filter. Instrument sensitivity is set to 75.
5. Readings taken from the Vero CCL-81 cells represent background levels of autofluorescence and are accordingly subtracted from the experimental measurements. After background subtraction, the fluorescence value obtained from the unintoxicated (control) Vero-d2EGFP cells is arbitrarily set at 100%. All experimental data are expressed as percentages of this 100% control value (see Note 17). Examples of the results produced are shown in Figs. 2a and 3.

4. Notes

1. See Chap. 6 in this volume for the use of cell-free bacterial culture supernatants in the Vero-d2EGFP assay.
2. In our experience, skilled high school and University undergraduates have been able to generate reproducible data with this system.
3. Apparently, Stx2 is no longer available from these vendors.
4. Any cell type could be stably transfected with d2EGFP to generate a reporter cell line for this assay. We identified Vero

cells as the most appropriate cell line for this work after several cell lines (Vero, BHK, CHO, Hep-2, HeLa, HEK 293, and COS-7) were characterized with dose-response curves to multiple plant and bacterial toxins. Of the cell lines tested, Vero cells displayed the highest level of sensitivity to a wide range of toxins.

5. When screening for toxin inhibitors, compounds such as proteasome inhibitors that block the turnover of d2EGFP produce a false positive result: in these instances, the d2EGFP signal persists despite the toxin-induced inhibition of protein synthesis. Screens for toxin inhibitors must therefore include a secondary assay to validate the hit compounds.
6. Transfection efficiency drops dramatically if the cells are grown for longer than 24 h in the six-well plate or if the cells are >80% confluent.
7. If more than one well of cells is to be transfected, the mixtures can be scaled up accordingly (i.e., three transfections would mix 3 μg of plasmid in 300 μL of DMEM for one microcentrifuge tube, and, in a separate tube, 15 μL of Lipofectamine would be mixed with another 300 μL of DMEM).
8. After identification, multiple plugs of the Vero-d2EGFP cell line are placed in freeze medium and stored in liquid nitrogen. No appreciable loss of fluorescence from our Vero-d2EGFP cell line has been detected over 5 months of continual passage.
9. For passaging cells, a 1:10 dilution from a confluent 10-cm dish takes 3–4 days to again reach confluency in a fresh 10-cm dish. A 1:20 dilution from a confluent 10-cm dish takes about 4–5 days to again reach confluency in a fresh 10-cm dish. Cells passaged at either dilution need fresh medium on the third day after transfer to a new dish.
10. Clear bottom, black-walled microplates are required to prevent well-to-well bleeding of the fluorescent signal.
11. Cell density influences toxin sensitivity, and highly confluent cells are more resistant to intoxication than subconfluent cells. For this reason, it is important to standardize from experiment to experiment the number of cells plated and the confluency of the cells at the time of toxin exposure.
12. The composition of various culture media produces a background fluorescent signal that can interfere with sample measurements. DMEM, for example, produces a strong fluorescent signal when using the 485/20 nm excitation filter and the 528/20 nm emission filter. To reduce this background signal, the culture medium is switched to Ham's F-12 during the

toxin incubation. Ham's F-12 medium produces a much weaker background fluorescence signal than DMEM. A 16–18 h incubation in serum-free Ham's F-12 medium does not affect the viability of Vero or Vero-d2EGFP cells.

13. Lab personnel who are learning the technique typically use 6 or 12 replicate wells for each experimental condition. After the procedure is established, three replicate wells can be used to generate consistent results. Toxin-free and toxin-containing media samples are distributed to the 96-well plate with a p200 pipette. The stated volumes can be scaled up in order to use pipette basins and a p200 multichannel pipette for sample distribution, but this leaves an excess of unused toxin at the end of the experiment.
14. A 16–18 h exposure to 100 μg cycloheximide/mL typically reduces the Vero-d2EGFP signal to about 10–20% of the maximal d2EGFP signal from unintoxicated control Vero-d2EGFP cells.
15. As the experiment is terminal at this point, the PBS does not need to be sterile. In addition, the processing steps can be performed at the bench rather than in a tissue culture hood.
16. To further reduce the background signal from cell culture media, final measurements are read from cells bathed in PBS.
17. When screening toxin inhibitors, each inhibitor should be applied to cells in the absence of toxin to establish a separate control (100%) value for that condition. This procedure corrects for variability that could result from inhibitor autofluorescence and/or inhibitor effects on cell viability in the absence of toxin.

Acknowledgments

The authors thank high school student Srikar Reddy and undergraduate Jessica Johnston for assistance with generating the dose-response curves for ricin, diphtheria toxin, and exotoxin A.

References

1. Rangel JM, Sparling PH, Crowe C, Griffin PM, Swerdlow DL (2005) Epidemiology of *Escherichia coli* O157:H7 outbreaks, United States, 1982–2002. Emerg Infect Dis 11: 603–609
2. Melton-Celsa AR, O'Brien AD (1998) Structure, biology, and relative toxicity of Shiga toxin family members for cells and animals. In: Kaper JB, O'Brien AD (eds) *Escherichia coli* O157:H7 and other Shiga

toxin-producing *E. coli* strains. ASM Press, Washington, DC, pp 121–128

3. Konowalchuk J, Speirs JI, Stavric S (1977) Vero response to a cytotoxin of *Escherichia coli*. Infect Immun 18:775–779
4. Paton JC, Paton AW (1998) Pathogenesis and diagnosis of Shiga toxin-producing *Escherichia coli* infections. Clin Microbiol Rev 11:450–479
5. Gamage SD, McGannon CM, Weiss AA (2004) *Escherichia coli* serogroup O107/O117 lipopolysaccharide binds and neutralizes Shiga toxin 2. J Bacteriol 186:5506–5512
6. Sekino T, Kiyokawa N, Taguchi T, Takenouchi H, Matsui J, Tang WR, Suzuki T, Nakajima H, Saito M, Ohmi K, Katagiri YU, Okita H, Nakao H, Takeda T, Fujimoto J (2004) Characterization of a shiga-toxin 1-resistant stock of vero cells. Microbiol Immunol 48:377–387
7. Hovde CJ, Calderwood SB, Mekalanos JJ, Collier RJ (1988) Evidence that glutamic acid 167 is an active-site residue of Shiga-like toxin I. Proc Natl Acad Sci U S A 85:2568–2572
8. Obrig TG, Louise CB, Lingwood CA, Boyd B, Barley-Maloney L, Daniel TO (1993) Endothelial heterogeneity in Shiga toxin receptors and responses. J Biol Chem 268:15484–15488
9. Zhao L, Haslam DB (2005) A quantitative and highly sensitive luciferase-based assay for bacterial toxins that inhibit protein synthesis. J Med Microbiol 54:1023–1030
10. Pastrana DV, FitzGerald DJ (2006) A nonradioactive, cell-free method for measuring protein synthesis inhibition by *Pseudomonas* exotoxin. Anal Biochem 353:266–271
11. Song S, Xue J, Fan K, Kou G, Zhou Q, Wang H, Guo Y (2005) Preparation and characterization of fusion protein truncated *Pseudomonas* Exotoxin A (PE38KDEL) in *Escherichia coli*. Protein Expr Purif 44:52–57
12. Quiñones B, Massey S, Friedman M, Swimley MS, Teter K (2009) Novel cell-based method to detect Shiga toxin 2 from *Escherichia coli* O157:H7 and inhibitors of toxin activity. Appl Environ Microbiol 75:1410–1416

Chapter 6

Use of a Vero Cell-Based Fluorescent Assay to Assess Relative Toxicities of Shiga Toxin 2 Subtypes from *Escherichia coli*

Beatriz Quiñones and Michelle S. Swimley

Abstract

Shiga toxin-producing *Escherichia coli* is a leading cause of human gastroenteritis from food and waterborne sources worldwide. Shiga toxins 1 and 2 are important virulence factors linked to severe human illness. In particular, Shiga toxin 2 is composed of a diverse and heterogeneous group of subtypes with differential cytotoxicities in mammalian cells. In this chapter, we describe the use of the Vero-d2EGFP fluorescent assay to examine the relative toxicities of Stx2 and Stx2 subtypes expressed by strains of Shiga toxin-producing *E. coli*.

Key words: Shiga toxin, Toxin detection, Foodborne pathogen, Food safety, Vero cells, Shiga toxin-producing *Escherichia coli*

1. Introduction

The enteric pathogen Shiga toxin-producing *Escherichia coli* (STEC) is known to cause human gastrointestinal illnesses, ranging from bloody diarrhea and hemorrhagic colitis to the life-threatening hemolytic uremic syndrome (HUS) (1, 2). The most commonly reported STEC serotype associated with large outbreaks and the development of HUS in North America is serotype O157:H7. Recently, findings from molecular typing studies have demonstrated that there is a prevalence of disease caused by other non-O157 serotypes, such as O26:H11, O103:H2, O111:H$^-$, O121:H19, and O145:H$^-$ in Europe and parts of Latin America, demonstrating that certain non-O157 STEC strains are potentially as virulent as O157:H7 strains (2, 3).

Otto Holst (ed.), *Microbial Toxins: Methods and Protocols*, Methods in Molecular Biology, vol. 739,
DOI 10.1007/978-1-61779-102-4_6, © Springer Science+Business Media, LLC 2011

One of the best characterized virulence factors in STEC pathogenicity and a determinant thought to be responsible for causing HUS are Shiga toxins (Stxs) (4, 5). Stxs are a class of AB_5 toxins, which consist of a catalytically active single A-subunit and a receptor-binding pentamer of B-subunits. The complete AB_5 ribosome-inactivating holotoxin is required for the inhibition of protein synthesis in human cells. Based on sequence similarity and catalytic activity, Stxs have been divided into two major groups: Stx1 and Stx2. Epidemiological studies suggest that STEC strains expressing Stx2 may be more virulent than strains expressing only Stx1 or both Stx1 and Stx2 (6–8).

In contrast to Stx1, the Stx2 group is composed of a diverse and heterogeneous group of subtypes (9). In addition to Stx2, different subtypes have been identified in STEC strains implicated in causing human illnesses; these subtypes are Stx2c (Stx2v-a), Stx2d (Stxd-OX3a and Stx2d-Ount), mucus-activatable Stx2d ($Stx2d_{activatable}$) (Stx2vh-a and Stx2vh-b), Stx2e, and Stx2f (10–14). Sequence analyses have shown that these Stx2 subtypes have a high sequence similarity to Stx2, and it is thought that these sequence variations may affect the ability of a particular strain to cause disease (8). Molecular typing studies have demonstrated that there is a strong correlation between STEC strains harboring certain *stx2* subtypes and severe disease outcomes such as bloody diarrhea and HUS (8, 15).

Recent evidence has shown that Stx2 and Stx2 subtypes appear to have different toxicities in cultured mammalian cells. One factor that may determine the Stx2-cytotoxic response is the receptor binding specificity of the target cells (16). The glycolipid globotriosylceramide (Gb_3) is the functional receptor for Stx2 and Stx2 subtypes (16, 17). Although globotetraosylceramide (Gb_4) was shown to be the preferred receptor for Stx2e (18), Gb_3 can substitute for Gb_4 in mediating the cytotoxic response of Stx2e (16, 19). In addition to receptor binding specificities, the amounts of Stx2 produced may define the severity of disease caused by STEC strains (20, 21). The differential expression and induction of Stx2 and Stx2 subtypes appear to contribute to the relative virulence of the STEC strain (22, 23).

To examine the relative toxicities of Stx2 and Stx2 subtypes expressed by STEC strains, the Vero-d2EGFP fluorescent assay was employed. The assay uses the Vero-d2EGFP cell line, generated from Vero cells to express a destabilized variant of the enhanced green fluorescent protein (EGFP) constitutively (24). The short, in vivo half-life of this EGFP variant makes it a sensitive marker for measuring the inhibition of protein synthesis by Stx. Vero cells are a suitable and sensitive system to examine Stx cytotoxicity since these cells contain large amounts of both Gb3 and Gb4 glycolipid receptors used by Stx2 and Stx2 subtypes (19). Thus, the Vero-d2EGFP fluorescent assay is a simple, highly sensitive, and quantitative method to examine the relative cytotoxicities of STEC strains.

2. Materials

2.1. Mammalian Cell Culture

1. Vero cells (ATCC CCL-81) (American Type Culture Collection, Manassas, VA).
2. Vero-d2EGFP cells, provided by Professor Ken Teter, University of Central Florida, Orlando, Florida, USA (see Note 1).
3. Antibiotic-antimycotic 100× solution containing 100 U/mL penicillin G sodium, 100 μg/mL streptomycin sulfate, and 25 μg/mL amphotericin B as Fungizone® in 0.85% saline (GIBCO, Grand Island, NY).
4. Dulbecco's modified Eagle's medium (DMEM)-high glucose with L-glutamine, and sodium pyruvate (GIBCO), supplemented with 10% of non-heat-inactivated, fetal bovine serum (American Type Culture Collection), and 1% antibiotic-antimycotic solution (GIBCO).
5. F-12 + GlutaMAX™-1 nutrient mixture (Ham's F-12) (GIBCO), supplemented with 10% of non-heat-inactivated, fetal bovine serum (American Type Culture Collection).
6. Solution of trypsin-EDTA (0.05%).
7. Fisher BioReagents® phosphate-buffered saline (PBS), 10× stock consisting of 1.37 M sodium chloride, 0.027 M potassium chloride, and 0.119 M phosphate buffer (Fisher Scientific, Fair Lawn, NJ). Prepare 1× working stock by diluting one part with nine parts of water, and sterilize by autoclaving.
8. Greiner Bio-One CELLSTAR® black 96-well polystyrene μClear® flat-bottom plates.
9. BD Falcon™ 75-cm^2 sterile tissue culture flasks with 0.2-μm vented plug seal caps.
10. Fisherbrand® 50-mL sterile pipette basins (Fisher Scientific).
11. Hausser Bright-line® hemocytometer with coverslip (Fisher Scientific).

2.2. Bacterial Culture

1. BD Falcon™ sterile and disposable round-bottom, polypropylene tubes with screw caps.
2. Luria-Bertani (LB) agar and broth.
3. Millex® polyvinylidene fluoride (Durapore®) syringe filters with a 0.45-μm pore size (Millipore Corporation, Bedford, MA).
4. Greiner Bio-One 96-well polypropylene microplates.

2.3. Fluorescence Microscopy

1. Nunc Lab-Tek® 8-well chamber slides with cover.
2. Fisherbrand® cover glass (22×22×0.1 mm) (Fisher Scientific).

3. Solution of paraformaldehyde prepared fresh for each experiment at 4% (w/v) in PBS.
4. Quench solution: 50 mM NH_4Cl in PBS.
5. Biømeda GEL/MOUNT™ aqueous mounting medium (Fisher Scientific).

3. Methods

The use of the Vero-d2EGFP cell line (see Note 1), generated from Vero cells, allows a sensitive and quantitative detection of active Stx. This detection method can be employed to examine the relative toxicities of Stx2 and Stx2 subtypes, expressed by various STEC strains (see Table 1). The use of epifluorescence and phase contrast microscopy allowed the visual inspection of the effects on the fluorescence of a subset of Vero-d2EGFP cells (see Fig. 1). The EGFP fluorescence was significantly diminished when Vero-d2EGFP cells were incubated with a 1,000-fold dilution of a cell-free, stationary-phase culture supernatant from Stx2- or Stx2d-expressing strains RM1918 and EH250, respectively. In contrast, higher EGFP fluorescence was observed after incubation with Stx2e-expressing strain S1191 or Stx2f-expressing strain T4/97.

A quantitative assessment of the Stx2-cytotoxic effects is carried out by measuring the average reduction of EGFP fluorescence from thousands of Vero-d2EFGP cells by using a 96-well plate reader. Dose-response curves are generated by incubating the Vero-d2EGFP cells for 16 h with various tenfold dilutions of cell-free supernatants, collected from STEC strains. Expression of Stx2 by RM1918 strain required a 10,000-fold dilution of culture supernatant to reach a 50% inhibition of EGFP fluorescence (see Fig. 2). This dilution of Stx2-containing culture supernatant was estimated to be two orders of magnitude more effective in inhibiting the EGFP signal when compared to the Stx2f-expressing strain T4/97, a 100 times more inhibitory than Stx2c- and Stx2e-expressing strains RM1625 and S1191, respectively, and approximately 50 times more inhibitory than Stx2d expressed by strain EH250. As a negative control, Vero-d2EGFP cells were treated with dilutions of a cell-free culture supernatant from a non-Stx-producing *E. coli* O157 strain RM4876; there was no loss of EGFP fluorescence for cells treated with supernatant from this non-Stx-producing strain.

3.1. Preparation of Crude Stx from STEC Culture Supernatants

1. STEC strains, each expressing distinct Stx2 and Stx2 subtypes (see Table 1), are in inoculated 5 mL sterile LB broth to a final OD_{600} = 0.2 (see Note 2).
2. Incubate the broths aerobically for 24 h at 37°C with orbital shaking at 200 rpm.

Table 1
***Escherichia coli* reference strains used in this study**

Strain	Shiga toxin type	Shiga toxin gene	Serotype	Source	Provider (Reference)[a]
RM1918	Stx2	stx_2/vt2	O157:H7	Human	R. E. Mandrell (25)
RM1625	Stx2c	stx_{2c}/vt2c	O157:H7	Cow	R. E. Mandrell (25)
EH250	Stx2d	stx_{-O118}/vt2-O118	O118:H12	Human	The STEC Center (14)
S1191	Stx2e	stx_{2e}/vt2e	O139:H1	Pig	R. E. Mandrell (26)
T4/97	Stx2f	stx_{2f}/vt2f	O128:H2	Pigeon	The STEC Center (10)
RM4876	None	None	O157:H7	Water	R. E. Mandrell (25)

[a]Affiliations of providers: R. E. Mandrell, US Department of Agriculture/ARS, Western Regional Research Center, Albany, CA; The STEC Center, National Food Safety and Toxicology Center, Michigan State University, East Lansing, MI

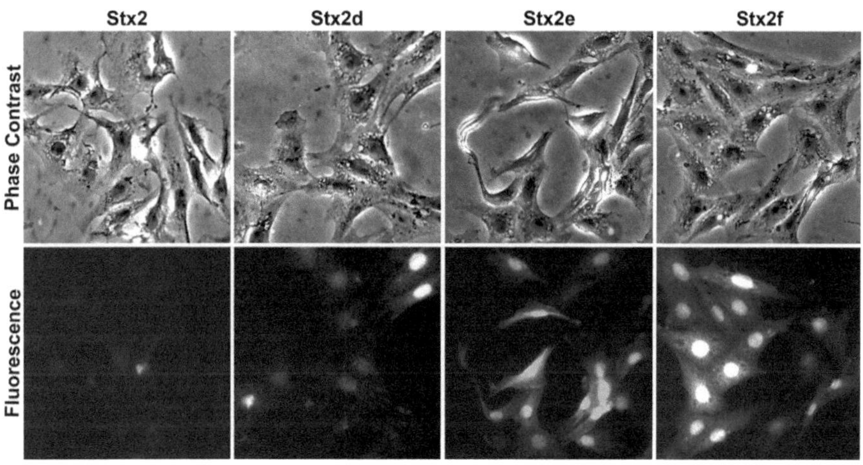

Fig. 1. Microscopic examination of Vero-d2EGFP cells treated with Stx2 and Stx2 subtypes. Digital phase contrast and fluorescent images of Vero-d2EGFP cells after incubation for 16 h in the presence of a 1,000-fold dilution of stationary-phase culture supernatant from Stx2-expressing *E. coli* O157:H7 strain RM1918, Stx2d-expressing *E. coli* O118:H12 strain EH250, Stx2e-expressing *E. coli* O139:H1 strain S1191, and Stx2f-expressing *E. coli* O128:H2 strain T4/97.

3. Pellet cells by centrifugation at 3,000 × *g* for 15 min.
4. Filter-sterilize STEC culture supernatants using 0.45-μm polyvinylidene fluoride syringe filters (see Note 3).
5. Store cell-free culture supernatants at 4°C for immediate use or at −20°C for long-term storage (see Note 4).

3.2. Treatment of Vero-d2EGFP Cells for Fluorescence Microscopy

1. The Vero-d2EGFP cells are passaged once monolayer has reached approximately 85% confluency on a 75-cm^2 tissue culture flask (see Note 5).
2. In a biological safety cabinet for cell culture, the DMEM is removed and the cells are washed twice in PBS.

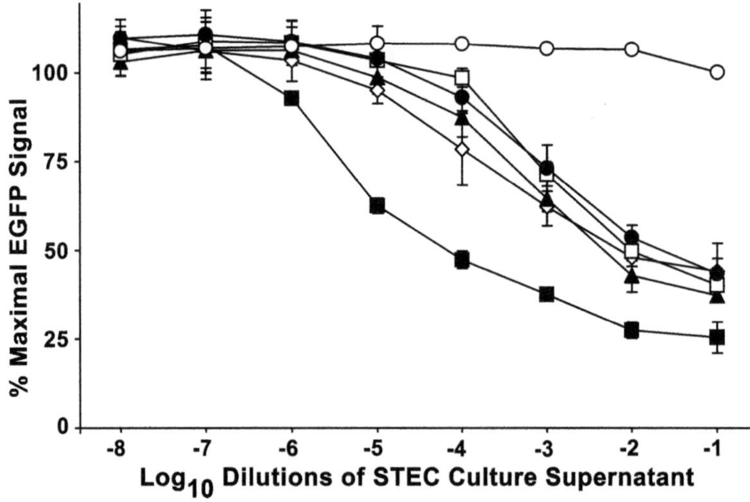

Fig. 2. Effect of Stx2 and Stx2 subtypes on the quantification of Vero-d2EGFP fluorescence. Vero-d2EGFP cells were exposed to Ham's F-12 medium containing several tenfold dilutions of cell-free culture supernatant from Stx2-expressing *E. coli* 0157:H7 strain RM1918 (*filled square*), Stx2c-expressing *E. coli* 0157:H7 strain RM6125 (*diamond*), Stx2d-expressing *E. coli* 0118:H12 strain EH250 (*open triangle*), Stx2e-expressing *E. coli* 0139:H1 strain S1191 (*open square*), Stx2f-expressing *E. coli* 0128:H2 strain T4/97 (*filled circle*), and non-Stx-expressing *E. coli* 0157:H7 strain RM4876 (*open circle*). Fluorescent signals were recorded with a microplate reader after a 16-h incubation period. The means ± standard deviation of three independent experiments with triplicate samples for each condition are shown.

3. Detach the cells from the flask by adding 1 mL trypsin-EDTA and incubate at 37°C for 5 min.
4. Resuspend the cells in 4 mL DMEM at a final volume of 5 mL.
5. Spin down the cells by centrifugation at 200 × *g* for 2 min.
6. Resuspend the cells in 10 mL DMEM.
7. By using a hemocytometer, prepare a Vero-d2EGFP cell suspension at 12,500 cells/mL.
8. Seed 5,000 Vero-d2EGFP cells/well (400 μl final volume/well) in sterile Lab-Tek® 8-well chamber slides with a Permanox® plastic surface (see Note 6).
9. Allow the cells to grow by incubating for 24 h at 37°C and 5% CO_2 under humidified conditions.
10. Remove DMEM and treat the cells with 400 μL of a 1:10 dilution of each STEC culture supernatant in DMEM. Incubate the treated cells at 37°C and 5% CO_2 under humidified conditions.
11. After 16-h incubation, remove the treatment media and rinse the intoxicated Vero-d2EGFP cells immediately twice with PBS.

12. To fix the cells, add paraformaldehyde solution to the cells for 20 min at 23°C with gentle shaking.
13. Remove paraformaldehyde and wash the cells three times in PBS.
14. Quench residual paraformaldehyde with 50 mM NH_4Cl for 10 min at 23°C.
15. Remove the quench solution and then quickly wash the cells twice with PBS.
16. Detach the slide from the media chamber by gripping the end of the slide with one hand. The gasket surrounding each well on the slide is removed using fine-tip tweezers. Add a drop of mounting medium on the cell preparation, mount coverslip, and seal the coverslip to the slide (see Note 7).
17. View the Vero-d2EGFP cells under phase contrast microscopy and then under fluorescence microscopy by using a DMR light microscope featuring an N PLAN 20× lens objective with a numerical aperture of 0.4 and a green fluorescent protein filter cube (Leica Microsystems, Inc., Bannockburn, IL). Digital phase contrast and fluorescence images are captured by using a Phoenix digital frame grabber (Active Silicon, Chelmsford, MA). An example of the results produced is shown in Fig. 1.

3.3. Treatment of Vero-d2EGFP Cells for Fluorescence Quantification

1. The Vero-d2EGFP and Vero cells are grown to approximately 85% confluency on a 75-cm^2 tissue culture flask.
2. The cells are detached from the flask by adding trypsin as described previously in the section above.
3. By using a hemocytometer, prepare a cell suspension at 90,000 cells/mL in Ham's F-12 medium (see Note 8).
4. Seed the Vero-d2EGFP cells at 9,000 cells/well (100 μl final volume) in a Greiner black 96-well microplate. To control for background, Vero cells are also seeded in Ham's F12 medium at 9,000 cells/well.
5. The cells are incubated for 24 h at 37°C in 5% CO_2 under humidified conditions.
6. Prepare several tenfold serial dilutions of STEC culture supernatants from each STEC strain in Ham's F-12 medium (see Note 9). Also prepare several tenfold serial dilutions of sterile LB without toxin in Ham's F-12 medium (see Note 10).
7. Treat the Vero-d2EGFP cells with 100 μL of each dilution of STEC culture supernatant.
8. Treat the Vero-d2EGFP and Vero cells with 100 μL of tenfold serial dilutions of LB broth without toxin in Ham's F12 medium as an unintoxicated control and as a control for background fluorescence, respectively.

9. Incubate the toxin-treated and untreated cells for 16 h at 37°C and 5% CO_2 under humidified conditions.
10. Remove the treatment medium from each well and then wash the cells in PBS for 3 min with gentle mixing. Repeat the PBS wash three more times (see Note 11).

3.4. Fluorescence Measurements and Data Analysis

1. EGFP fluorescence is measured on a Synergy™ HT multi-detection microplate reader (BioTek, Winooski, VT) with the 485/20-nm excitation filter and the 528/20-nm emission filter.
2. For each assay, the sensitivity of the photomultiplier tube is set to 75, and the bottom probe optics position is selected for taking the measurements when using the Synergy™ microplate reader (see Note 12).
3. Fluorescent signals from Vero-d2EGFP cells are background corrected by subtracting the fluorescence recorded for the control Vero cells.
4. Results for the toxin-treated Vero-d2EGFP cells are expressed as a percentage of the values obtained for control cells incubated without toxin. An example of the results produced is shown in Fig. 2.

4. Notes

1. See Chap. 5 in this volume for the description on the generation of the Vero-d2EGFP cell line.
2. Strains are grown in LB rather than in tryptic soy broth, a common growth medium for *E. coli*. In our experience, tryptic soy broth exhibits higher levels of autofluorescence than LB.
3. Filter units with polyvinylidene fluoride membranes (Durapore® membranes, Millipore Corp.) are one of the lowest protein-binding membranes available and are the optimal membranes for filtering STEC culture supernatants for crude toxin preparations. Avoid using filters with mixed cellulose esters or hydrophilic nylon membranes since these membranes exhibit protein binding.
4. To minimize repetitive freeze-thawing of crude toxin preparation from STEC culture supernatants, aliquots in microcentrifuge tubes are prepared and stored at –20°C. Gentle mixing of supernatants is recommended, and vortexing should be avoided.
5. For passaging Vero-d2EGFP and Vero cells, a 1:10 dilution from a confluent 75-cm^2 flask will provide an experimental culture that will approach confluence after 72 h.

6. Vero-d2EGFP cells attach well to both Permanox® plastic and glass growth surfaces that are available for Lab-Tek® chamber slides.
7. To avoid undesirable air bubbles in the mounting medium, lower the coverslip slowing and carefully over the sample. Remove any excess mounting media by pressing gently with a tissue wipe before sealing the coverslip with nail varnish.
8. See Note 12 of Chap. 5 in this volume on the use of cell culture media DMEM versus Ham's F-12 when treating Vero-d2EGFP cells.
9. Depending on the levels of expression of Stx2 and Stx2 subtypes by a particular STEC strain, it may be necessary to prepare more than 8 tenfold serial dilutions of the culture supernatants to determine the half-maximal effective concentration.
10. Each tenfold dilution per condition is tested in triplicate, and each experiment is repeated three times. Vero cells must be seeded on each experimental 96-well plate to control for background signal. Efficiency can be increased with the use of 8- or 12-channel pipettes with 96-well polypropylene plates (low-protein binding) when preparing dilutions of cell-free supernatants containing toxin.
11. LB broth exhibits some autofluorescence; therefore, it is important to wash the treated Vero-d2EGFP cells multiple times in PBS to minimize this background signal.
12. The sensitivity of the photomultiplier tube may have to be adjusted for some assays to ensure that the fluorescent signal falls within the dynamic range of the plate reader. Setting the optics position to take fluorescence measurements with the bottom probe provides better results when working with adherent mammalian cells.

Acknowledgments

The authors wish to thank Professor Ken Teter from the University of Central Florida, Orlando, Florida, for providing the mammalian cell lines Vero-d2EGFP and Vero, and Robert E. Mandrell from the USDA/ARS/Western Regional Research Center, Albany, California, for providing the STEC strains used in this study. This work was supported by the US Department of Agriculture, Agricultural Research Service CRIS project number 5325-42000-045.

References

1. Rangel JM, Sparling PH, Crowe C, Griffin PM, Swerdlow DL (2005) Epidemiology of *Escherichia coli* O157:H7 outbreaks, United States, 1982–2002. Emerg Infect Dis 11:603–609
2. Karmali MA (2004) Infection by Shiga toxin-producing *Escherichia coli*: an overview. Mol Biotechnol 26:117–122
3. Bettelheim KA (2007) The non-O157 Shiga-toxigenic (verocytotoxigenic) *Escherichia coli*; under-rated pathogens. Crit Rev Microbiol 33:67–87
4. O'Brien AD, Kaper JD (1998) Shiga Toxin-Producing *Escherichia coli*: yesterday, today, and tomorrow. In: Kaper JD, O'Brien AD (eds) Escherichia coli O157:H7 and other Shiga Toxin-producing E. coli strains. ASM Press, Washington, DC, pp 1–11
5. Serna A, Boedeker EC (2008) Pathogenesis and treatment of Shiga toxin-producing *Escherichia coli* infections. Curr Opin Gastroenterol 24:38–47
6. Boerlin P, McEwen SA, Boerlin-Petzold F, Wilson JB, Johnson RP, Gyles CL (1999) Associations between virulence factors of Shiga toxin-producing *Escherichia coli* and disease in humans. J Clin Microbiol 37:497–503
7. Ostroff SM, Tarr PI, Neill MA, Lewis JH, Hargrett-Bean N, Kobayashi JM (1989) Toxin genotypes and plasmid profiles as determinants of systemic sequelae in *Escherichia coli* O157:H7 infections. J Infect Dis 160:994–998
8. Müthing J, Schweppe CH, Karch H, Friedrich AW (2009) Shiga toxins, glycosphingolipid diversity, and endothelial cell injury. Thromb Haemost 101:252–264
9. Scheutz F, Strockbine NA (2005) Genus I. *Escherichia*. In: Garrity GM, Brenner DJ, Krieg NR, Staley JT (eds) Bergey's manual of systematic bacteriology, 2nd edn. Springer, New York, pp 607–624
10. Schmidt H, Scheef J, Morabito S, Caprioli A, Wieler LH, Karch H (2000) A new Shiga toxin 2 variant (Stx2f) from *Escherichia coli* isolated from pigeons. Appl Environ Microbiol 66:1205–1208
11. Schmitt CK, McKee ML, O'Brien AD (1991) Two copies of Shiga-like toxin II-related genes common in enterohemorrhagic *Escherichia coli* strains are responsible for the antigenic heterogeneity of the O157:H- strain E32511. Infect Immun 59:1065–1073
12. Ito H, Terai A, Kurazono H, Takeda Y, Nishibuchi M (1990) Cloning and nucleotide sequencing of Vero toxin 2 variant genes from *Escherichia coli* O91:H21 isolated from a patient with the hemolytic uremic syndrome. Microb Pathog 8:47–60
13. Melton-Celsa AR, Kokai-Kun JF, O'Brien AD (2002) Activation of Shiga toxin type 2d (Stx2d) by elastase involves cleavage of the C-terminal two amino acids of the A_2 peptide in the context of the appropriate B pentamer. Mol Microbiol 43:207–215
14. Piérard D, Muyldermans G, Moriau L, Stevens D, Lauwers S (1998) Identification of new verocytotoxin type 2 variant B-subunit genes in human and animal *Escherichia coli* isolates. J Clin Microbiol 36:3317–3322
15. Persson S, Olsen KE, Ethelberg S, Scheutz F (2007) Subtyping method for *Escherichia coli* Shiga toxin (verocytotoxin) 2 variants and correlations to clinical manifestations. J Clin Microbiol 45:2020–2024
16. Samuel JE, Perera LP, Ward S, O'Brien AD, Ginsburg V, Krivan HC (1990) Comparison of the glycolipid receptor specificities of Shiga-like toxin type II and Shiga-like toxin type II variants. Infect Immun 58:611–618
17. Waddell T, Head S, Petric M, Cohen A, Lingwood C (1988) Globotriosyl ceramide is specifically recognized by the *Escherichia coli* verocytotoxin 2. Biochem Biophys Res Commun 152:674–679
18. DeGrandis S, Law H, Brunton J, Gyles C, Lingwood CA (1989) Globotetraosylceramide is recognized by the pig edema disease toxin. J Biol Chem 264:12520–12525
19. Keusch GT, Jacewicz M, Acheson DWK, Donohue-Rolfe A, Kane AV, McCluer RH (1995) Globotriaosylceramide, Gb3, is an alternative functional receptor for Shiga-like toxin 2e. Infect Immun 63:1138–1141
20. Dean-Nystrom EA, Melton-Celsa AR, Pohlenz JFL, Moon HW, O'Brien AD (2003) Comparative pathogenicity of *Escherichia coli* O157 and intimin-negative non-O157 Shiga toxin-producing *E. coli* strains in neonatal pigs. Infect Immun 71:6526–6533
21. Zhang X, McDaniel AD, Wolf LE, Keusch GT, Waldor MK, Acheson DWK (2000) Quinolone antibiotics induce Shiga toxin-encoding bacteriophages, toxin production, and death in mice. J Infect Dis 181:664–670
22. de Sablet T, Bertin Y, Vareille M, Girardeau JP, Garrivier A, Gobert AP, Martin C (2008) Differential expression of stx2 variants in Shiga toxin-producing *Escherichia coli* belonging to seropathotypes A and C. Microbiology 154:176–186

23. Zhang W, Bielaszewska M, Friedrich AW, Kuczius T, Karch H (2005) Transcriptional analysis of genes encoding Shiga toxin 2 and its variants in *Escherichia coli*. Appl Environ Microbiol 71:558–561
24. Quiñones B, Massey S, Friedman M, Swimley MS, Teter K (2009) Novel cell-based method to detect Shiga toxin 2 from *Escherichia coli* O157:H7 and inhibitors of toxin activity. Appl Environ Microbiol 75: 1410–1416
25. Cooley M, Carychao D, Crawford-Miksza L, Jay MT, Myers C, Rose C, Keys C, Farrar J, Mandrell RE (2007) Incidence and tracking of *Escherichia coli* O157:H7 in a major produce production region in California. PLoS ONE 2:e1159
26. Marques LRM, Peiris JSM, Cryz SJ, O'Brien AD (1987) *Escherichia coli* strains isolated from pigs with edema disease produce a variant of Shiga-like toxin II. FEMS Microbiol Lett 44:33–38

Chapter 7

Molecular Methods: Chip Assay and Quantitative Real-Time PCR: In Detecting Hepatotoxic Cyanobacteria

Anne Rantala-Ylinen, Hanna Sipari, and Kaarina Sivonen

Abstract

Cyanobacterial mass occurrences are widespread and often contain hepatotoxic, i.e. microcystin- and nodularin-producing, species. Nowadays, detection of microcystin (*mcy*) and nodularin synthetase (*nda*) genes is widely used for the recognition of toxic cyanobacterial strains in environmental water samples. Chip assay presented here combines ligation detection reaction and hybridization on a universal microarray to detect and identify the *mcyE/ndaF* genes of five cyanobacterial genera specifically and sensitively. Thus, one chip assay can reveal the co-occurrence of several hepatotoxin producers. The presented quantitative real-time PCR method is used for the detection of either microcystin-producing *Anabaena* or *Microcystis*. Determination of the *mcyE*-gene copy numbers allows the identification of the dominant producer genus in the sample.

Key words: Cyanobacteria, Microcystin/nodularin synthetase genes, Chip, qPCR, Molecular detection methods

1. Introduction

Cyanobacterial mass occurrences (blooms) are a world-wide phenomenon. Blooms pose a risk for water users, since they often contain toxins. The most common cyanobacterial toxins are the hepatotoxins: microcystins and nodularins are cyclic peptides that are structurally similar. Microcystins have seven amino acids and nodularins five, respectively (1). Toxic effects in mammals are due to the inhibition of protein phosphatases 1 and 2A (2) and are restricted mostly to liver cells, where toxin molecules enter via the bile salt transport system (3). The main producers of microcystins are planktic, freshwater strains of the genera *Microcystis*, *Planktothrix*, and *Anabaena*. Nodularins are produced solely by *Nodularia spumigena* strains in brackish waters (1).

Otto Holst (ed.), *Microbial Toxins: Methods and Protocols*, Methods in Molecular Biology, vol. 739, DOI 10.1007/978-1-61779-102-4_7, © Springer Science+Business Media, LLC 2011

Microcystins and nodularins are produced non-ribosomally by microcystin and nodularin synthetases, respectively. These large biosynthetic enzyme complexes are comprised of non-ribosomal peptide synthetases, polyketide synthases, mixed peptide–polyketide synthetases, and modifying enzymes. The enzymes are encoded by the microcystin (*mcy*) and nodularin synthetase (*nda*) genes (4).

Cyanobacterial mass occurrences can contain many cyanobacterial species, none, one, or several of which may include toxin-producing strains (1, 5, 6). However, microcystin-producing strains appear similar to nontoxic ones and cannot be recognized with conventional microscopy (1), thus making the assessment of the risk for water users difficult. Genes encoding the synthetase enzymes have instead been proven useful in the differentiation and identification of hepatotoxin producers (4).

The chip assay protocol presented here uses ligation detection reaction (LDR) and universal microarray (7) to detect and identify simultaneously all potential microcystin and nodularin producers present in samples (8, 9). The recognition is based on the PCR amplification step prior to LDR. DNA extracted from environmental samples is amplified with the so-called general primers that target *mcyE/ndaF* genes of all the main producer genera (5, 8, 10). For successful amplification, DNA extracted from an environmental sample needs to be purified and should not contain impurities that inhibit the PCR polymerase. In addition, PCR primers and conditions should allow an unbiased amplification of the different *mcyE/ndaF* genes. In LDR, the amplified *mcyE/ndaF* genes of different genera are recognized by two genus-specific probes (Fig. 1). The discriminating probe has a fluorescent dye (Cy3) at its 5′ end. At its 3′ end, the discriminating probe contains nucleotides unique to the target gene sequence. The common probe is phosphorylated at the 5′ end and has a complementary ZipCode (cZipCode) sequence at its 3′ end. Ligation requires adjacent attachment of the discriminating and common probes to their target as well as perfect base pairing at the probe junction. As a result, the two probes form a single molecule with a Cy3 at one end and a cZipCode at the other. During successive LDR cycles, newly formed probe molecules are released from and new probes attached to the target sequences, amplifying the signal originating from the different *mcyE/ndaF* genes present. In hybridization, cZipCodes find their corresponding ZipCodes that are printed as spots on glass slides. The cZipCodes and probes that annealed to their target gene and ligated during LDR also carry the Cy3 dye to the spots. Fluorescence emitted from the spots is detected by laser scanner and indicates the presence of the corresponding *mcyE/ndaF* genes in the sample. The chip assay can be extended to detect cyanobacterial groups present in samples by the addition of the 16S rRNA gene PCR and

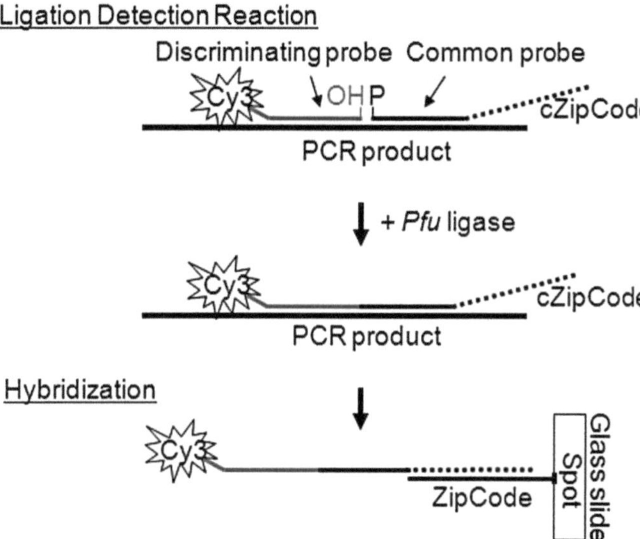

Fig. 1. Principle of the chip assay. In the first step of ligation detection reaction (LDR), discriminating and common probes find their target PCR products (=amplified *mcyE* genes of different producer genera). If both probes attach one after each other, they are ligated together. As a result, a fluorescent dye (Cy3) and a address sequence (cZipCode) are joined in the same molecule that is released from the target in the next step of the cyclic LDR. The LDR steps are then repeated to amplify the fluorescent signals. In hybridization, the fluorescent signals are addressed to the spots on a glass slide via recognition of the corresponding ZipCodes by the cZipCodes [Figure modified from Ref. 14].

group-specific probes for different 16S rRNA genes in the protocol (9, 11).

Quantitative real-time PCR (qPCR) method of this protocol uses primers and TaqMan probes that specifically amplify and detect *mcyE* genes of either *Anabaena* or *Microcystis* (9). In addition to the identification of the producer/producers, qPCR allows for quantification of gene copies present in samples and, thus, reveals the dominant producer genus. Quantity of the *mcyE* genes in samples is interpolated from a standard curve (12) that is drawn based on the amplification of a standard dilution series prepared from genomic DNA of an *Anabaena* or a *Microcystis* strain (9).

2. Materials

2.1. PCR of the mcyE/ndaF Genes

1. Primers: mcyE-F2 (5′-GAAATTTGTGTAGAAGGTGC-3′) or mcyE-F2b (5′- TGAAATTTGTGTAGAAGGTG-3′) and mcyE-R4 (5′-AATTCTAAAGCCCAAAGACG-3′) (see Note 1); dissolve the primers with sterile deionized water to 100 μM (=100 pmol/μL; stock concentration). For use in PCR, dilute an aliquot of the stock to 20 pmol/μL. Store both at −20°C.

2. Super Taq plus polymerase (5 U/µL) and 10× Super Taq plus buffer (HT Biotechnology Ltd). Store at –20°C.
3. Nucleotide solution (dNTP mix, Finnzymes; concentration of each nucleotide 10 mM). Store at –20°C.
4. Bovine serum albumin (BSA) acetylated (10 mg/mL, Promega). Store at –20°C.
5. Template DNA extracted from environmental samples. Use 30–40 ng for a PCR. Store at –20°C.
6. DNA of a microcystin- or nodularin-producing strain to be used as positive control sample. Store at –20°C.
7. Sterile deionized water, used also as a negative control sample.
8. PCR tubes or strips.
9. PCR machine.
10. 50× Tris/acetic acid/EDTA (TAE) buffer (Bio-Rad). Mix 10 mL with 990 mL of Milli-Q water to have 1,000 mL of ready-to-use buffer (0.5× TAE).
11. Agarose (LE, analytical grade, V3121 Promega).
12. Ethidium bromide (EtBr) solution (10 mg/mL, Bio-Rad).
13. Any size marker that can be used to estimate if the length of the amplicons is correct (≈810 bp), e.g. GeneRuler 100-bp Plus DNA ladder (Fermentas).
14. 6× DNA loading dye (Fermentas); supplied with the size marker.
15. Wide Mini-Sub Cell GT for running agarose gels (Bio-Rad).
16. PowerPac 300 power supply (Bio-Rad).
17. E.Z.N.A. Cycle-Pure kit (Omega Biotek), ethanol (min. 96 vol%).

2.2. Ligation Detection Reaction

1. Oligomix: Includes discriminating probes with a Cy3 label at 5′ ends (see Note 2), and common probes phosphorylated at 5′ ends and cZipCode sequences attached to the 3′ ends (probes detecting *mcyE/ndaF* genes according to Ref. 8; probes for 16S rRNA genes according to Ref. 11). Dilute and mix probes into a solution in which each probe is at 250 fmol/µL concentration. Store at –20°C.
2. LDR control oligonucleotide, 5′-AGCCGCGAACACCACGATCGACCGGCGCGCGCAGCTGCAGCTTGCTCATG-3′. Dilute to concentration of 25 fmol/µL. Store as aliquots at –20°C.
3. *Pfu* DNA ligase (4 U/µL) and 10× *Pfu* DNA ligase buffer (Stratagene, Agilent Technologies). Store at –20°C.

2.3. Hybridization

1. Microarray slides with ZipCodes that correspond to the probe pairs and cZipCodes of oligomix (see Note 3).
2. Saline-sodium citrate (SSC) buffer (20× concentrate; Sigma). Store at 20–22°C (room temperature).
3. Sodium dodecyl sulfate (SDS, 10% solution; Sigma). Store at room temperature.
4. Albumin from bovine serum ≥96% (BSA; Sigma). Store at 2–8°C.
5. Pre-hybridization solution (5× SSC, 1% BSA): Measure 0.5 g of BSA in a 50-mL plastic tube. Add approximately 20 mL of water and dissolve BSA, e.g. by vortexing. Add 12.5 mL of 20× SSC and fill with water to 50 mL and mix. Before filling the tube, decrease the amount of foam formed during mixing by centrifugation. Store at 4°C if prepared on the previous day.
6. Wash solution I (1× SSC, 0.1% SDS): Prepare in a 1-L bottle. Mix 50 mL of 20× SSC, 10 mL of 10% SDS, and 940 mL of distilled water to get 1,000 mL of ready-to-use solution. Store at room temperature.
7. Wash solution II (0.1× SSC): Prepare in a 1-L bottle. Mix 5 mL of 20× SSC, and 995 mL distilled water to get 1,000 mL of ready-to-use solution. Store at room temperature.
8. Salmon testes DNA (10 mg/mL; Sigma). Store at −20°C.
9. Hybridization control: 5′-Cy3-GTTACCGCTGGTGCT GCCGCCGTA-3′ (8). Dissolve and dilute to a concentration of 10 fmol/µL. Store as aliquots at −20°C.
10. Hybridization chamber system: The protocol uses a custom-made hybridization chamber with Press-to-Seal silicone isolator (1.0×9 mm; Schleicher and Schuell Bioscience, Dassel, Germany) to create chambers for simultaneous hybridization of eight samples (Fig. 2).
11. Temperature-controlled water bath/baths.

2.4. Signal Detection and Image Analysis

1. GenePix Autoloader 4200AL laser scanner (Axon Instruments, Inc).
2. GenePix Pro Microarray Acquisition and Analysis Software for GenePix microarray scanners (Axon Instruments, Inc).

2.5. qPCR of Anabaena- and Microcystis-mcyE Genes

1. *Anabaena-mcyE*specific primers: 611F (5′-CTAGAGTAGT CACTCACGTC-3′) and 737R (5′-GGTTCTTGATAGTTA GATTGAGC-3′), and a probe: 672P (5′-CAAGTTCCCA CAATTCTTGGATTAGCAGC-3′) (9); *Microcystis-mcyE* specific primers: 127F (5′-AAGCAAACTGCTCCCGG TATC-3′) and 247R (5′-CAATGGGAGCATAACGAGT CAA-3′), and a probe: 186P (5′-CAATGGTTATCGAATT GACCCCGGAGAAAT-3′) (9). Dissolve primers and probes with sterile deionized water to 100 µM (=100 pmol/µL; stock

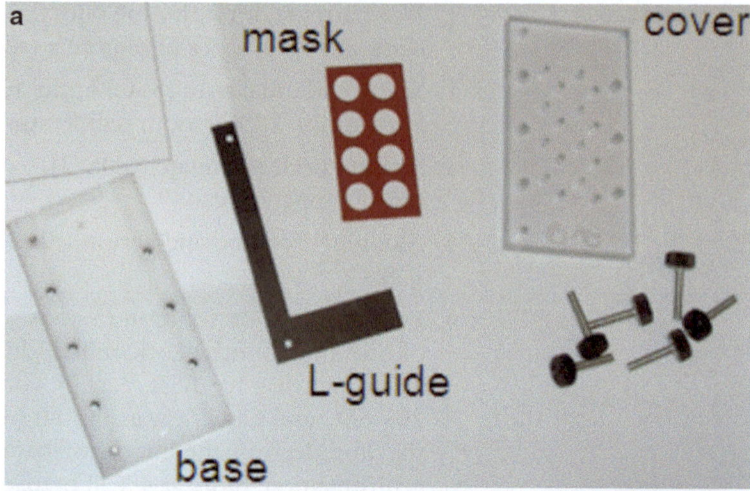

Fig. 2. Hybridization chamber used in the protocol. (**a**) The chamber includes a base with a slit to place the glass (not in the figure) in the right position. The L-shaped guide is used to orientate the red rubber mask in order to align the individual chambers with printed arrays. The L-guide is removed before adding the cover, which is fastened with six screws. Both the L-guide and cover are aligned with the base using two metallic pins on the left-hand corners of the base and the corresponding holes in the guide and cover. (**b**) The hybridization chamber is placed in a plastic box with water-saturated tissues on the bottom. This creates a moist climate in the box and prevents hybridization mixture from drying-out during hybridization.

concentration). For use in qPCR, dilute an aliquot of the stock to a concentration of 10 pmol/μL. Store both at –20°C.

2. TaqMan® Universal PCR Master Mix (Applied Biosystems) or iTaq™ Supermix with ROX Sample (Bio-Rad), and sterile deionized water.
3. ABI PRISM 96-well optical reaction plates, ABI PRISM optical adhesive covers, and MicroAmp splash-free support base (Applied Biosystems).

4. ABI7300 instrument (Applied Biosystems).
5. Genomic DNA from microcystin-producing *Anabaena* and *Microcystis* strain for external standard dilution series.

3. Methods

3.1. PCR of the mcyE/ndaF Genes

1. Let dNTPs, 10× buffer, BSA, polymerase, primers, and environmental DNA samples melt on ice. In the meanwhile, mark the PCR tubes or strips and program the following protocol in the PCR machine: 95°C, 3 min; 30 cycles of (94°C, 30 s; 53°C, 30 s; 68°C, 1 min); 68°C, 10 min; 4°C, ∞.
2. Prepare the PCR mix for eight environmental DNA samples (two replicates of each), a positive, and a negative (water) control sample. In addition, reserve some extra mix to compensate for pipetting losses. For example, 20 reactions, add in a 1.5-mL tube (keep on ice), 236 μL of sterile deionized water, 40 μL of Super Taq plus buffer, 50 μL of BSA, 10 μL of dNTPs, 10 μL of both mcyE-F2/mcyE-F2b and mcyE-R4, and 4 μL Super Taq plus polymerase. Mix well by vortexing. Pipette 18 μL of the mix in each 0.2-mL PCR tube. Finally, add 2 μL of samples and controls into PCR tubes to have a total volume of 20 μL. Close the lids carefully and spin down if needed.
3. Run the PCR program detailed above.
4. During the PCR run, prepare a 1.5% agarose gel. For a 100-mL gel, weigh 1.5 g of agarose and pour into an Erlenmayer flask. Add 100 mL of 0.5× TAE buffer and swirl to loosen the agarose off the bottom. Heat the agarose in a microwave oven until all the agarose is melted. If the solution threatens to boil over, stop the microwave, swirl the flask carefully, and continue heating. Use protective gloves, because the flask can be extremely hot. Add 1 μL of EtBr in a solution that has been cooled down to approximately 55–60°C. Mix well and pour to the gel cast and insert the comb. Remove all air bubbles. Let solidify for 0.5–1 h.
5. After PCR, combine the replicated reactions. Run 5 μL of each reaction (mixed with 1 μL of loading dye) in the agarose gel. Add size marker to one of the wells. Run the gel at 100 V for 30 min.
6. Inspect the gel under UV light. Remember to protect your eyes and skin.
7. If amplification was successful based on gel image, i.e. one band of ca. 810 bp is visible in each band, purify the rest of the combined eight PCRs with E.Z.N.A. Cycle-Pure Kit

according to the instructions provided with the kit. Elute DNA with 50 μL of sterile water.

8. Measure DNA concentration with a spectrophotometer. The concentration can be calculated (if not provided by the device on hand) using the OD value at 260 nm wavelength: $A_{260} \times 50$ μg/mL × dilution factor of the sample × sample volume (mL).

3.2. LDR

1. For LDR, 25 fmol (=0.025 pmol) of PCR products is used. The corresponding amount in nanograms is 13.4 ng and is calculated as follows: 0.025 pmol (PCR product) × length of the amplified PCR product (810 bp) × mass of one base pair (=660 pg/pmol) × 1 ng/10^3 pg. Divide 13.4 ng by the DNA concentration of PCR product to find out how many microliters of PCR product is needed for LDR. If the volume needed is <2 μL, dilute the PCR product prior to preparing the LDR to increase the accuracy.
2. For one LDR, add sterile water (volume depends on the amount of PCR product), 2 μL of 10× *Pfu* ligase buffer, 1 μL of oligomix, 1 μL of LDR control oligo, and X μL of PCR product, to have a total volume of 19 μL. Use PCR tubes or strips. Keep the tubes on ice.
3. Close caps, heat the reactions for 2 min at 94°C (use a PCR machine), spin, and place the tubes back on ice.
4. Add 1 μL of *Pfu* DNA ligase to each reaction. Close caps and spin again.
5. Cycle in a PCR machine with the following program: 94°C, 2 min; 30 × (90°C, 30 s; 63°C, 4 min); 94°C, 2 min; 4°C, 10 min.
6. Store at −20°C until use.

3.3. Hybridization

1. Prepare pre-hybridization solution. Warm the water bath to 42°C and let the pre hybridization solution warm in the bath. After the temperature is reached, set the microarray glass into the tube (see Note 4). Incubate for 1 h in the dark.
2. After incubation, wash the glass with distilled water five times for 30 s. Dry by centrifugation at 200 × *g* for 5 min.
3. Increase the temperature of the water bath to 65°C.
4. Set up the hybridization chamber system (Fig. 2). Tighten the screws carefully to prevent leakage from one hybridization chamber to another.
5. Prepare the hybridization mix for eight samples (plus one extra to compensate for pipetting losses). For nine reactions, add 243 μL of sterile deionized water, 146.7 μL of 20× SSC,

6.3 μL of salmon testes DNA, and 9 μL of hybridization control in a 1.5-mL tube and mix by vortexing.

6. Add 45 μL of the hybridization mix into eight LDR tubes (á 20 μL). Incubate at 94°C for 2 min (use a PCR machine) and chill on ice. Spin the tubes. Transfer the mixtures (á 65 μL) to individual hybridization chambers. Avoid the formation of air bubbles that prevent annealing between cZipCodes and ZipCodes.
7. Place the hybridization chamber system in a plastic box (with water-saturated tissues in the bottom; Fig. 2b) in a water bath (65°C) and incubate for 2 h in dark. Warm up 50 mL of washing solution I in the water bath during hybridization. Seal the tube carefully.
8. During hybridization, prepare the other washing solutions. Pour washing solution II (1×) and water (3×) in 50-mL tubes. Reserve approximately 100 mL of washing solution I (room temperature) for opening the hybridization chamber.
9. Open the hybridization chamber submerged in washing solution I (room temperature) to avoid drying the hybridization mix on the hot glass, which would cause high background signal level. Transfer the glass slide in the pre-warmed washing solution I using forceps, seal tightly, and place back on the water bath. Incubate at 65°C for 15 min. Transfer the slide to washing solution II. Incubate for 5 min at room temperature. Finally, wash the slide in Milli-Q water three times for 5 min. During each wash, mix a couple of times by inverting the tube.
10. Dry the slide by centrifugation at 200 ×*g* for 5 min.
11. Store at room temperature in dark. Preferably, proceed to scanning immediately or on the following day at the latest.

3.4. Signal Detection

1. First, switch on the microarray scanner and then the computer. Open the image acquisition program. For detailed instructions, refer to *User's guide and tutorial* of the program. Wait until the program tells it is ready for scanning and loading the slides.
2. Load the glass slide on the slide carrier tray so that the printed side faces you and the bottom of the slide is on the left. Load the tray in the scanner.
3. Open "hardware settings" window (button available on the right-hand side of the main window), select wavelength 532 nm (deselect the other wavelengths), and set PMT gain to 500 and laser power (%) to 100.
4. Press prescan icon (►►) to scan the whole slide with a 40-μm resolution.

5. From the resulting image, select an area (=one of the eight arrays). In the hardware settings, set “Pixel size” to 5 μm and “Lines to average” to four. Press scan icon (▶) to scan the selected area.
6. The fluorescent spots should be green. Any white spots might indicate saturated signal level (signal value over 65535). Lower the PMT gain value until the spots are green.
7. Save the image in a 16-bit TIFF format using the command *Save images* in the *File* menu (button on the right-hand side of the main window).
8. Repeat prescan and scan steps for the remaining seven arrays.

3.5. Image Analysis

1. These instructions assume that the image analysis is performed separately from the image acquisition using another computer. For detailed instructions, refer to the *User's guide and tutorial* of the GenePix Pro program.
2. Open the image analysis program (analysis-only mode). Select *Open images* in the *File* menu.
3. Import a GenePix Array List (GAL) file using the *Load Array List* command. A GAL file contains data on the arrayed spots (=features), e.g. size, position, and identity, and based on the data, creates a block with circular feature-indicators on an image.
4. Place the block roughly over the image. Under *Align Blocks* (button on the left side of the Image tab), select *Align Features in Selected Blocks*. This command aligns feature-indicators with spots. If any of the clearly visible spots is not found, align its feature-indicator manually. First, right click the mouse and select *Feature Mode* from the menu. Select a feature-indicator by clicking on it. To move and resize a feature-indicator, use the arrow keys and [Ctrl]+arrow keys, respectively. To select previous or next feature-indicator, use [Alt]+arrow keys.
5. After aligning the feature-indicators, perform the analysis by pressing the *Analyze* button on the right-hand side of the window.
6. Click *Save Results As* in the *File* menu, check the *Save a JPEG image* check box, and save. This creates a JPEG figure (Fig. 3) of the scanned image and a GPR file with the analyzed data.
7. Click *Export Results* in the *File* menu to export the data in a text format.
8. Use Excel to open the text file and to evaluate the signals from spots. Each gene is represented by four spots in an array. To be sure that a certain *mcyE/ndaF* gene was present in the sample, signals of the spots corresponding to it need to fulfill

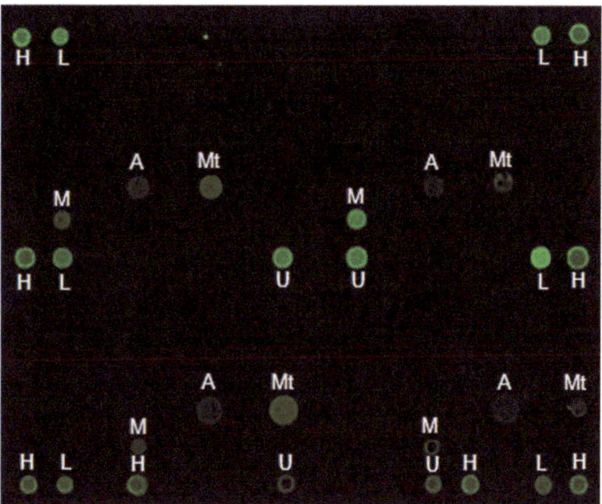

Fig. 3. A scanned image of an array. Target of the chip assay was a water sample from Lake Tuusulanjärvi bloom. In addition to eight hybridization (H) and six LDR control (L) spots, signals are detected from four spots corresponding to 16S rRNA genes of *Anabaena/Aphanizomenon* (A) and *Microcystis* (M) as well as the *mcyE* genes of *Microcystis* (Mt). U stands for spots corresponding to probe pairs recognizing any cyanobacterial 16S rRNA gene.

three conditions: (I) signal-to-noise ratio (SNR) of a spot is greater than or equal to three, (II) over 70% of the spot's signal intensity is two standard deviations higher than the background signal intensity (column *% > B532+2SD* in the results sheet), and (III) at least two of the replicated spots fulfill the first and second condition. In Excel, use the filter tool (Data/Filter/AutoFilter/Custom…) to select spots that meet these criteria.

9. The signals need to be normalized in order to compare signals originating from a certain gene in different hybridizations or arrays. To do this, the average gene-specific (four replicates) signal intensity (column *F532 Median – B532* in the results sheet) is presented as a percentage of average signal intensity of either hybridization (eight replicates) or LDR (six replicates) control spots.

3.6. qPCR of Anabaena- and Microcystis-mcyE Genes

1. Prepare a standard dilution series containing 10^1, 10^2, 10^3, 10^4, 10^5, and 10^6 copies of *Anabaena-mcyE* or *Microcystis-mcyE* gene, using genomic DNA of microcystin-producing *Anabaena* or *Microcystis*, respectively. Calculate first the mass of one genome using the following equation: genome size (bp) $\times$ 660 g/mol (=approximate mass of one base pair)/6.022045×10^{23} bp/mol = genome size (bp) $\times 1.096 \times 10^{-21}$ g/bp. If 5.8 Mb and 7.1 Mb genome sizes are assumed for *Microcystis* (13) and *Anabaena* (http://www.ncbi.nlm.nih.gov/genomes/lproks.cgi), masses of their genomes are 6.36 fg and 7.78 fg, respectively. Since one genome contains one copy of the *mcyE* gene,

this amount of DNA also contains one copy of the *mcyE* gene. For 10^6 copies of the gene, 6.36 and 7.78 ng of DNA in one qPCR are needed. Taking into account the concentrations of genomic DNA extractions, make dilutions to have a concentration of 6.36/7.78 ng/5 μL (=volume pipetted into one reaction, see Note 5). Prepare enough dilution for three replicate reactions and for preparing the 10^5 dilution, e.g. to 5 μL of the 10^6 dilution add 45 μl of sterile water. Prepare the other dilutions in a similar way. Prepare the standard dilution series just before performing qPCR. Do not freeze-thaw the dilutions. Keep on ice.

2. Make 1:100 (see Note 6) dilutions of the environmental DNA samples. Prepare enough for three replicate qPCRs. Keep on ice.
3. Switch on the computer and the ABI7300 machine. After the PCR machine is ready, turn on the 7,300 System SDS program. Fill in the sample data for the run: sample name, identity (unknown/standard/negative control), and standard sample's quantity. Remember to add detector to each well that is used. If the detector is not added, no data will be gathered during the run. Program the following thermal profile: 50°C, 2 min; 95°C, 10 min; 40 × (95°C, 15 s; 62°C, 1 min). Change the sample volume to 25 μL (default value is 50 μL). Select data collection to happen at "Stage 3, Step 2 (62.0 @ 1:00)."
4. Calculate the total number of qPCRs (=samples, standard dilutions, and negative control, all in three replicates) and prepare PCR mix in a 1.5-mL or 2.0-mL tube. For a 25-μL reaction, pipette 12.5 μL of TaqMan Universal PCR Master Mix (or iTaq Supermix), 5.5 μL of sterile water, 0.75 μL of both primers, and 0.5 μL of TaqMan probe. Prepare some extra PCR mix to cover for pipetting losses. The amount depends on the number of samples; the more samples, the more extra mix needed. Place the 96-well plate in a support base to keep the bottom of the plate clean and to avoid transferring any contaminating substances into PCR block. Dispense 20 μL of the mix to each of the wells. Then, add 5 μL of samples, standard dilutions, and negative control (sterile water). Seal the plate with an adhesive cover. Do not touch the cover with bare hands.
5. Run the PCR program.
6. After run, analyze the amplification plot with the following settings: *Manual Ct* = 0.2 and *Manual Baseline Start (cycle)* = 3 and *End (cycle)* = 15. In the results sheet, *Ct* values of each amplification as well as mean quantity and standard deviation of gene copy numbers are given.
7. Export the results in CSV format (File/Export/Results…) and open the file in Excel to handle the data more easily.

8. The copy numbers in the results sheet refer to copy numbers in a reaction. To estimate the number of gene copies in the original sample, take into account the dilution factor (100), volume of the extracted DNA (e.g. 100 μL), and volume of the water sample (e.g. 200 mL) used for DNA extraction. For example, if a quantity of 200 copies was detected in a reaction (5 μL of 1:100 dilution), then 20,000 copies were in 5 μL of the original DNA; 400,000 copies (=20,000 × 100 μL/5 μL) were in the whole DNA extraction; and 2,000 copies (=400,000/200 mL) were present in 1 mL of the original water sample.

4. Notes

1. Primer pair mcyE-F2b/R4 amplifies the *ndaF* gene of nodularin-producing *Nodularia spumigena* more efficiently than mcyE-F2/R4 and should be used if samples are from brackish waters. Both pairs are equally suitable for studying samples containing microcystin producers.
2. Minimize exposure to light in all the steps of protocol to prevent loss of fluorescence.
3. Amino-modified ZipCode oligonucleotides have previously been printed on the CodeLink slides (Amersham) in 100 mM phophate buffer, pH 8.5 (11), or on the Nexterion® A slides (Schott Nexterion) in 1× Micro Spotting Solution Plus (MSP, TeleChem International) (9).
4. Store glass slides protected from light and dust. Use gloves and forceps, when handling slides. Avoid touching/scratching the printed area.
5. Using small volumes (<2 μL) increases inaccuracy of pipetting and causes greater proportional variation between replicates.
6. Correct dilution factor depends on samples. For reliable quantification, amplification of the unknown samples should start (=exceed the threshold cycle; see Ref. 12) within the limits set by the standard curve. If amplification starts too late or too early, use a smaller or a bigger dilution factor, respectively.

Acknowledgments

This work was supported by grants from the Academy of Finland to A.R.-Y. (128480) and K.S. (Research Center of Excellence 53305, 118637, and the grant for Academy Professors 214457).

References

1. Sivonen K, Jones G (1999) Cyanobacterial toxins. In: Chorus I, Bartram J (eds) Toxic cyanobacteria in water: a guide to their public health consequences, monitoring, and management. E & FN Spon, London, UK
2. MacKintosh C, Beattie KA, Klumpp S, Cohen P, Codd GA (1990) Cyanobacterial microcystin-LR is a potent and specific inhibitor of protein phosphatases 1 and 2A from both mammals and higher plants. FEBS Lett 264:187–192
3. Carmichael WW (1994) The toxins of cyanobacteria. Sci Am 270:78–86
4. Sivonen K, Börner T (2008) Bioactive compounds produced by cyanobacteria. In: Herrero A, Flores E (eds) The cyanobacteria: Molecular biology, genomics and evolution. Caister Academic Press, Norfolk, UK
5. Rantala A, Rajaniemi-Wacklin P, Lyra C, Lepistö L, Rintala J, Mankiewicz-Boczek J, Sivonen K (2006) Detection of microcystin-producing cyanobacteria in Finnish lakes with genus-specific microcystin synthetase gene E (*mcyE*) PCR and associations with environmental factors. Appl Environ Microbiol 72:6101–6110
6. Fewer DP, Köykkä M, Halinen K, Jokela J, Lyra C, Sivonen K (2009) Culture-independent evidence for the persistent presence and genetic diversity of microcystin-producing *Anabaena* (Cyanobacteria) in the Gulf of Finland. Environ Microbiol 11:855–866
7. Gerry NP, Witowski NE, Day J, Hammer RP, Barany G, Barany F (1999) Universal DNA microarray method for multiplex detection of low abundance point mutations. J Mol Biol 292:251–262
8. Rantala A, Rizzi E, Castiglioni B, de Bellis G, Sivonen K (2008) Identification of hepatotoxin-producing cyanobacteria by DNA-chip. Environ Microbiol 10:653–664
9. Sipari H, Rantala-Ylinen A, Jokela J, Oksanen I, Sivonen K (2010) Development of a chip assay and quantitative PCR for detecting microcystin synthetase E gene expression. Appl Environ Microbiol 76:3797–3805
10. Rantala A, Fewer DP, Hisbergues M, Rouhiainen L, Vaitomaa J, Börner T, Sivonen K (2004) Phylogenetic evidence for the early evolution of microcystin synthesis. Proc Nat Acad Sci USA 101:568–573
11. Castiglioni B, Rizzi E, Frosini A, Sivonen K, Rajaniemi P, Rantala A, Mugnai MA, Ventura S, Wilmotte A, Boutte C, Grubisic S, Balthasart P, Consolandi C, Bordoni R, Mezzelani A, Battaglia C, De Bellis G (2004) Development of a universal microarray based on the ligation detection reaction and 16S rRNA gene polymorphism to target diversity of cyanobacteria. Appl Environ Microbiol 70:7161–7172
12. Applied Biosystems, Real-Time PCR Systems, Part #: 4348358 Rev E., Chemistry Guide. Available at: www3.appliedbiosystems.com/AB_Home/Support/Tutorials Troubleshooting/index.htm#)
13. Kaneko T, Nakajima N, Okamoto S, Suzuki I, Tanabe Y, Tamaoki M, Nakamura Y, Kasai F, Watanabe A, Kawashima K, Kishida Y, Ono K, Shimizu Y, Takahashi C, Minami C, Fujishiro T, Kohara M, Katoh M, Nakazaki N, Nakayama S, Yamada M, Tabata S, Watanabe MM (2007) Complete genomic structure of the bloom-forming toxic cyanobacterium *Microcystis aeruginosa* NIES-843. DNA Res 14:247–256
14. Rantala A (2007) Evolution and detection of cyanobacterial hepatotoxin synthetase genes. PhD thesis, Edita Prima Oy, Helsinki, Finland (available at: http://urn.fi/URN:ISBN:978-952-10-4369-7)

Part II

Endotoxins

Chapter 8

Capillary Electrophoresis Chips for Fingerprinting Endotoxin Chemotypes from Whole-Cell Lysates

Béla Kocsis, Anikó Kilár, Lilla Makszin, Krisztina Kovács, and Ferenc Kilár

Abstract

Endotoxins (lipopolysaccharides, LPSs) are components of the envelope of Gram-negative bacteria. These molecules, responsible for both advantageous and harmful biological activities of these microorganisms, are highly immunogenic and directly involved in numerous bacterial diseases in humans such as Gram-negative sepsis. The characterization of endotoxins is of importance, since their physiological and pathophysiological effects depend on their chemical structure. The differences among LPSs from different bacterial serotypes and their mutants include variations mainly within the composition and length of their O-specific polysaccharide chains.

Proper assignation of the *S* or *R* chemotypes of endotoxins is possible by analyzing their electrophoretic profiles. The recent microchip electrophoretic methods provide fast characterizations and differentiations of endotoxins. The methods are applicable for determination directly from whole-cell lysates after destruction of the proteinaceous components by proteinase K digestion and precipitation of the LPS components. The partially purified LPS components are visualized either by interaction with dodecyl sulfate and a fluorescent dye, or by a covalently bound fluorescent dye. These chip electrophoretic methods have advantages of high speed and quantification and replace the sodium dodecyl sulfate-polyacrylamide gel electrophoresis with silver staining.

Key words: Microchip electrophoresis, Whole-cell lysate, Endotoxin, Lipopolysaccharide, LPS, Chemotype, Fluorescent labeling

1. Introduction

Endotoxins (lipopolysaccharides, LPSs) are components of the envelope of Gram-negative bacteria. These molecules, responsible for both advantageous and harmful biological activities of these microorganisms, are highly immunogenic and directly involved in numerous bacterial diseases in humans such as Gram-negative sepsis (1). The amphiphilic LPS compounds consist of a hydrophobic

Otto Holst (ed.), *Microbial Toxins: Methods and Protocols*, Methods in Molecular Biology, vol. 739,
DOI 10.1007/978-1-61779-102-4_8,

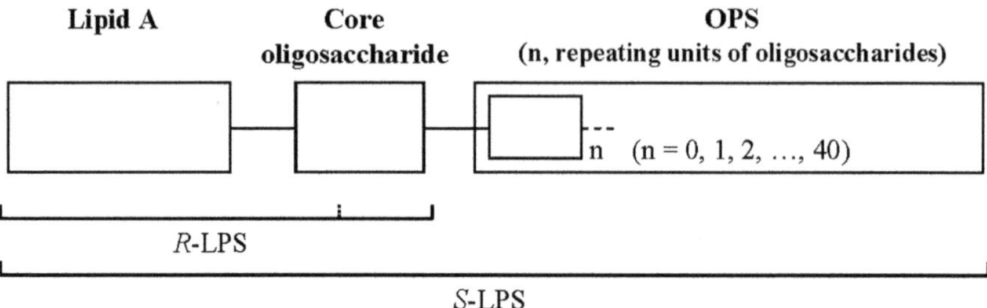

Fig. 1. Basic structure of endotoxins (*LPSs*, lipopolysaccharides) from Gram-negative bacteria (*OPS*, O-specific polysaccharide).

lipid region (lipid A) covalently linked to a hydrophilic oligo- or polysaccharide region (2). Typically, the polysaccharide region of smooth LPS (*S*-LPS) chemotypes consists of a core and an O-specific polysaccharide (OPS), whereas the rough LPS (*R*-LPS) chemotypes lack this and sometimes even core components (see Fig. 1). The differences among the *S*-LPSs from different bacterial serotypes and their mutants include variations mainly within the composition and length of the OPSs. The characterization of endotoxins is of importance, since their physiological and pathophysiological effects depend on their chemical structure.

When a large number of bacterial mutants and their LPS content are to be compared, for instance, in the preparation of vaccines, LPSs are detected directly from bacterial cultures, generally by the relatively quick method described by Hitchcock and Brown (3). This procedure involves enzymatic digestion and lysis of whole cells, leading to partially purified, protein-free total cellular LPS. An important specialty of this method is that only 1 mL culture volume is necessary for the characterization of LPSs, besides having a short overall process time (ca. 40 h).

The analysis of intact LPS molecules and the classification of the chemotypes are routinely checked by sodium dodecyl sulfate-polyacrylamide gel electrophoresis (SDS-PAGE) with silver staining (4). With this method, 18 LPS samples can be analyzed in a Bio-Rad Miniprotean system within ca. 24 h. The presence of bands near the front line corresponds to the simpler *R*-type structures, while a distinctive ladder-like pattern reflects *S*-type structures containing OPSs with an increasing number of repeating units.

The microchip electrophoretic method to fingerprint bacterial LPSs from whole-cell lysate samples is based on LPS complexation with dodecyl sulfate and a fluorescent dye (5), or with a covalently bound fluorescent dye (6). In the first method, the LPSs are disaggregated by SDS resulting in LPS–SDS complexes, which are then visualized with a fluorescent dye having strong tendency to create complexes with SDS (7). In the other method, the fluorescent dye is covalently bound to the LPSs. The labeled endotoxin complexes are analyzed in microchips (in practice, using the Agilent 2100 bioanalyzer microchip electrophoresis system) applying the *Agilent Protein 80 LabChip* kit or the *High Sensitivity Protein 250 LabChip* kit, respectively. The separation is performed with a sieving matrix included in the kits, which provides optimal resolution for the separation of LPS molecules with different compositions, chain lengths, and concentration. With the use of these microchip analyses, 11 endotoxin assignations can be performed within ca. 1 h.

2. Materials

2.1. Bacterial Cell Culture and Lysis

1. Bacterial stock is stored at –80°C.
2. 5 mL of Mueller-Hinton broth (Oxoid Ltd., UK): 3.0 g beef extract, 17.5 g acid hydrolysate of casein, and 1.5 g starch in 1 L deionized water. The broth is stored at 4°C.
3. Lysozyme is dissolved in water at 100 mg/mL and stored at -20°C.
4. LPS lysing buffer: 2% (w/v) SDS, 4% (v/v) β-mercaptoethanol, 10% (w/v) glycerol, 1 M Tris–HCl buffer, pH 6.8, and 0.05 % (w/v) bromophenol blue. Store at 25°C.
5. Proteinase K is stored at –20°C.
6. Ethanol/magnesium chloride solution: magnesium chloride (3.8 mg) is dissolved in 50 mL ethanol and stored at 25°C. The reagents are of analytical grade.

2.2. Lipopolysaccharide–Dodecyl Sulfate–Fluorescent Dye Complexes

1. Sieving matrix: a polydimethylacrylamide-based linear polymer in an SDS-containing buffer at pH 8.0 (this gel matrix is included in the *Agilent Protein 80 Labchip* kit) is mixed thoroughly (with Vortex) at 3.85% with a fluorescent dye named Agilent dye concentrate (included in the *Agilent Protein 80 Labchip* kit and has 630 nm as excitation and 650 nm as emission wavelengths) until an uniform color is obtained. Store at 4°C and protect the reagent from light (see Note 1). As the dye binds to nucleic acids, it should be treated as a potential mutagen, besides that it is dissolved in dimethyl sulfoxide (DMSO), which is known to facilitate the entry of organic molecules into tissues; therefore, it should be used with appropriate care.
2. Destaining solution: the composition of this solution is the same as that of the sieving matrix (included in the *Agilent Protein 80 Labchip* kit), except it does not contain the fluorescent dye. Store at 4°C.
3. SDS solution at 4% (w/v) is stored at 25°C.

2.3. Lipopolysaccharides Covalently Labeled with Fluorescent Dye

1. 30 mM Tris/HCl buffer at pH 8.5 is stored at 4°C.
2. Sieving matrix: a polydimethylacrylamide-based linear polymer in an SDS-containing buffer at pH 8.0 (this is included in the *Agilent High Sensitivity Protein 250 Labchip* kit).
3. Destaining solution: this solution is included in the *Agilent High Sensitivity Protein 250 LabChip* kit.
4. Sample buffer/DTT solution: the sample buffer is included in the *Agilent High Sensitivity Protein 250 LabChip kit* and is mixed with 3.5% (v/v) 1 M 1,4-dithiothreitol (DTT). Store at –20°C.
5. Labeling dye: this is a fluorescent dye included in the *Agilent High Sensitivity Protein 250 LabChip* kit and mixed with DMSO as following: 54 μL DMSO is added onto the pellet of 1 vial of labeling dye and mixed until all solid components are completely dissolved (the fluorescent dye reagent needs reconstitution in DMSO, which is known to facilitate the entry of organic molecules into tissues; therefore, it should be used with appropriate care). Store the dye protecting it from light (see Note 1).
6. Ethanolamine (this is corrosive) is stored at –20°C.

2.4. Electrophoresis on Microchip

1. Microchip developed for protein sizing (7) (e.g. Agilent Protein chip from Agilent Technologies).
2. Agilent 2100 bioanalyzer system equipped with a diode laser for fluorescence detection (Agilent Technologies, Waldbronn, Germany) and with 2100 Expert software.
3. Syringe priming station for filling the microchip (Agilent Technologies).
4. Electrode cleaner (Agilent Technologies). This is a microchip without capillaries, thus the space connecting the wells can be filled with ca. 350 μL of aqueous solution.

3. Methods

These instructions describe the analyses of partially purified endotoxin samples from whole-cell lysates of bacteria, and assume the use of the Agilent 2100 bioanalyzer microchip electrophoresis system. The methods are easily adaptable to other microchip formats. It is critical that the molecular mass range of the sieving matrix used for the separation of LPS components is optimal and that appropriate fluorescent dye is used for labeling and detection.

3.1. Preparation of LPS Samples from Whole-Cell Lysate

1. Bacterial stock are streaked on Mueller-Hinton agar plate and grown for 18 h at 37°C in a bacterial incubator.
2. One colony is transferred into 5 mL of culture medium (Mueller-Hinton broth) using a sterile inoculating loop and incubated at 37°C on a shaker for 18 h.

3. 1 mL of the cell culture is collected in an Eppendorf tube and washed with 1 mL of water once by centrifugation at 6,000 × *g* for 3 min.
4. The pellets are resuspended in 1 mL of water and heated at 100°C for 30 min.
5. 200 μL of the cooled suspension is taken out in an Eppendorf tube and 4 μL of the lysozyme solution is added. The mixture is heated at 37°C for 30 min. During this process, the peptidoglycan layers of the bacterial cell wall are disintegrated.
6. To the cooled mixture, 200 μL of the LPS lysing buffer is added and the lysates are incubated at 100°C for 10 min. During this step, the cellular materials, such as proteins, LPSs, nucleic acids, and cell debris, are released from the bacterial cells.
7. For proteolytic digestion, proteinase K enzyme is dissolved at 20 μg/μL in water. 20 μL of this enzyme solution is added in two 10-μL portions to the cell lysate and incubated at 65°C for 4 h (the second portion is added at the middle of the incubation time, i.e. after 2 h).
8. The LPS content is precipitated by adding 800 μL volume of the ethanol/magnesium chloride solution and the mixture is stored at −20°C for 18 h.
9. The mixture is centrifuged at 13,000 × *g* for 15 min, then the sediment (containing LPS) is suspended in 30 μL of deionized water, and sonicated in an ultrasonic bath.

3.2. Preparation of Lipopolysaccharide–Dodecyl Sulfate–Fluorescent Dye Complexes

1. 4-μL portions of whole-cell lysate LPS samples are mixed with 2 μL of 4% SDS solution (see Note 2).
2. The sample mixtures are sonicated in water bath, incubated at 100°C for 5 min, centrifuged (6,000 × *g*, 15 s), and diluted five times with deionized water.
3. 12 μL of the sieving matrix is pipetted (see Note 3) in the well marked “ ” and the chip channels are filled hydrodynamically (the Agilent Protein chip needs a special syringe priming station for the filling, which is described in Subheading 3.4, step 1). After the filling, make sure that no bubbles are present in the capillaries. The filling is proper if the capillaries cannot be seen on the backside of the chip.
4. 12 μL of the sieving matrix is pipetted (see Note 3) in the other three wells labeled with “G” and 12 μL of the destaining solution is pipetted in the well marked “DS” (see Note 4).
5. 6 μL of whole-cell lysate LPS samples is loaded on the sample wells in the microchip (see Note 3), as well as in the ladder well marked (see Note 5). All sample wells and the ladder well of the microchip must be filled before electrophoretic analysis.
6. The loaded microchip is placed in the Agilent 2100 bioanalyzer equipment (see Note 6) (for the following steps, see Subheading 3.4 of step 2).

3.3. Preparation of Lipopolysaccharides Covalently Labeled with Fluorescent Dye

1. The whole-cell lysate LPS samples are added to the Tris/HCl buffer in 9:1 ratio (see Note 2), e.g. 9 μL LPS sample and 1 μL Tris/HCl buffer.
2. 10 μL of the whole-cell lysate LPS samples in Tris buffer is mixed with 0.5 μL of the fluorescent dye/DMSO solution and incubated for 30 min on ice (see Note 2).
3. 1 μL of ethanolamine is added and incubated for 10 min on ice. During this step, the excess dye is quenched after reaction by ethanolamine.
4. The whole-cell lysate LPS sample mixtures are diluted tenfold with deionized water.
5. 4 μL of the labeled and diluted whole-cell lysate LPS sample mixtures is combined with 2 μL of the sample buffer/DTT solution and incubated at 100°C for 5 min, and then centrifuged (6,000 × *g*, 15 s).
6. For loading the wells on the microchip, see Subheading 3.2, steps 3–6.

3.4. Electrophoresis on Microchip Using Agilent 2100 Bioanalyzer

1. Filling the capillaries of the microchip with sieving matrix: the base plate of the syringe priming station is adjusted to position "A" and the syringe clip is adjusted to its middle position; the microchip is put in the priming station and a plunger is put at 1 mL; the syringe priming station is closed and the plunger is pressed until held by the clip; the clip is released after 90 s and after 5 s, the plunger is slowly pulled back to 1-mL position; and the priming station is opened and the remaining wells are filled with the proper solutions (see Subheading 3.2, steps 4–5).
2. The loaded chip is placed in the receptacle of the Agilent 2100 bioanalyzer and the lid is carefully closed so that the electrodes in the cartridge fit into the wells of the chip.
3. The appropriate assay (Protein 80 Series II.xsy for the analysis of the non-covalent LPS–SDS–dye complexes and High Sensitivity Protein 250.xsy assay for the analysis of the covalent LPS–dye complexes) is selected from the *Assay* menu and the assay is started immediately by clicking on the start button (see Note 7).
4. The analyses of all samples in the microchip are accomplished within 30 min (see Note 8). The data procession by the software includes baseline correction, alignment of separated runs in one chip, estimation for molecular masses (see Note 5), and the display of the peak data in the microchip electropherograms as gel-like images on a gray scale (this is not shown here). In all samples, the peaks originating from the LPS components appear directly after the relatively big peak corresponding to

the SDS–dye complex (see Note 9) when analyzing LPS–SDS–dye complexes, or to the ethanolamine–fluorescent dye complex when analyzing LPS–dye complexes (see Note 10). The *S*-LPSs appear as multiple "waves" of peaks (LPS components) with a broad molecular mass distribution (containing both high and low mobility components), according to the number of repeating units in the OPS, while the *R*-LPSs contain only one (or two) peak(s) with high mobility (low molecular mass) component(s) since they lack the OPS. Examples of results are shown in Figs. 2 and 3.

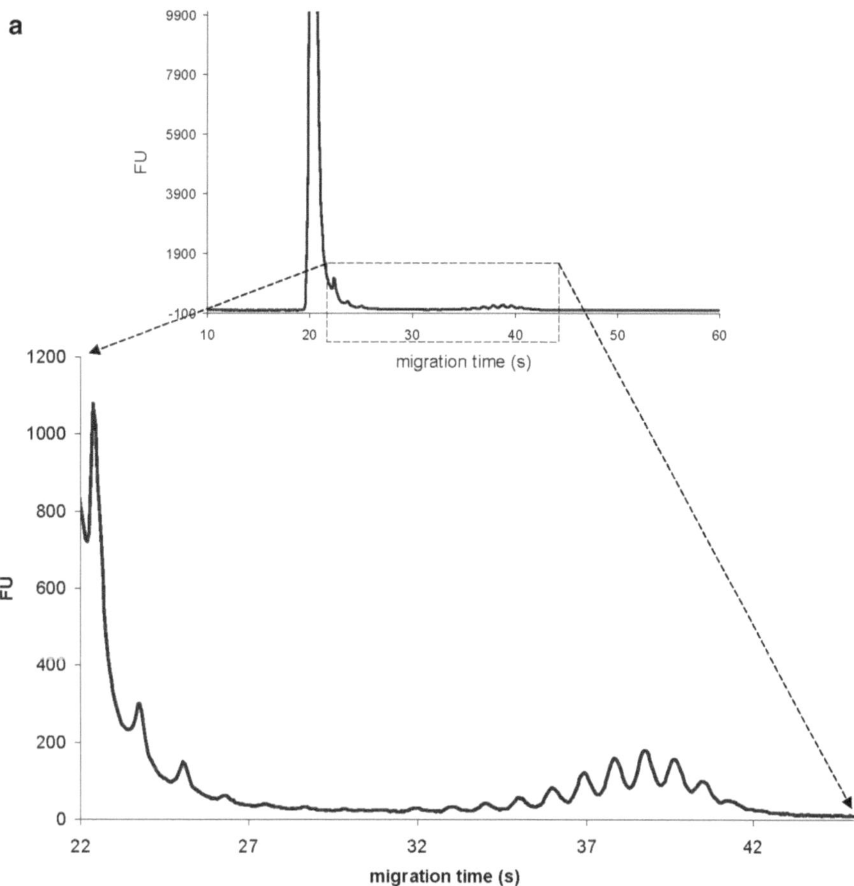

Fig. 2. Microchip electrophoretic profiles of partially purified *S*-LPS sample from *Escherichia coli* 083 whole-cell lysate. The LPS molecules are visualized (**a**) by covalently bound fluorescent dye employing the *High Sensitivity Protein 250 LabChip* kit or (**b**) in the form of non-covalent endotoxin–dodecyl sulfate–fluorescent dye complexes employing the *Agilent Protein 80 LabChip* kit. Equipment: Agilent 2100 bioanalyzer with laser-induced fluorescence detector. The microchip (Agilent Protein chip) is filled with a polydimethylacrylamide-based linear polymer of high viscosity (its adsorption to the capillary wall reduces electroendosmotic flow to almost zero, thus the negatively charged LPS–dye or LPS–SDS–dye complexes migrate toward the anode, i.e. the positive pole). The first large peaks in the inserts correspond to the system peaks (signals of the non-bound, noncomplexed dye), while the subsequent peaks show the LPS components having 0, 1, 2, etc., repeating units in the OPS (enlarged electropherograms).

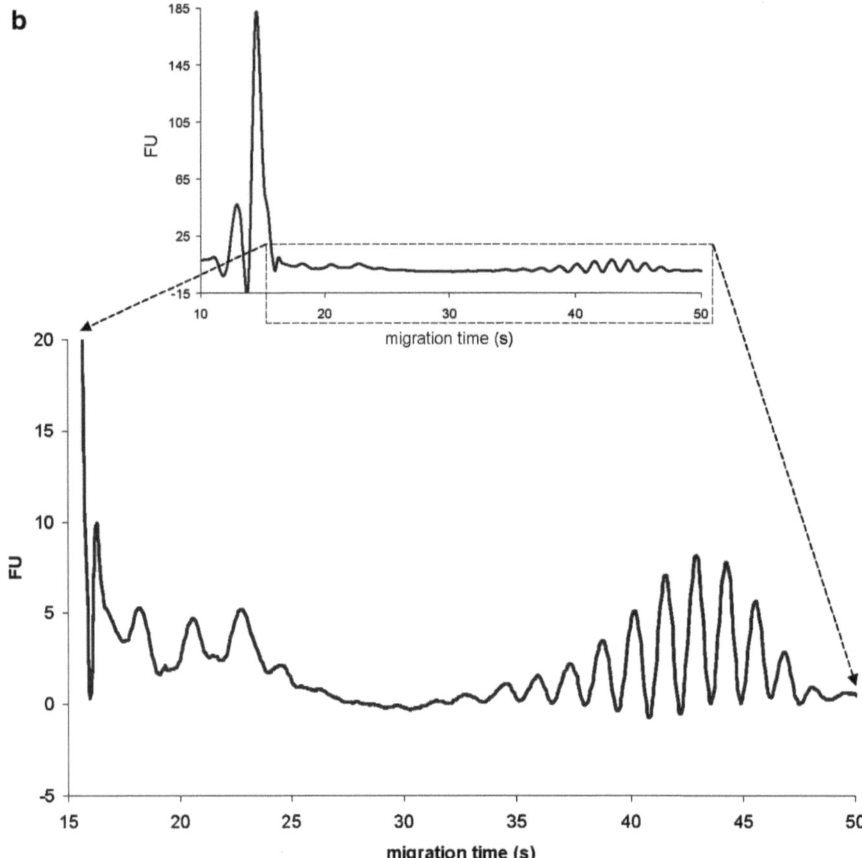

Fig. 2. (continued)

5. The chip is immediately removed from the Agilent 2100 bioanalyzer when the assay is completed (see Note 11).
6. One of the wells of the electrode cleaner is slowly filled with 350 µL of deionized water and the electrode cleaner is placed in the Agilent 2100 bioanalyzer and left for about 10 s after closing the lid, to remove possible contaminants from the electrodes.

4. Notes

1. Remove light covers only when pipetting. The dye contained in the reagents decomposes when exposed to light and this reduces the signal intensity.
2. Use 0.5 mL vials. Using larger vials may lead to poor results, caused by evaporation.

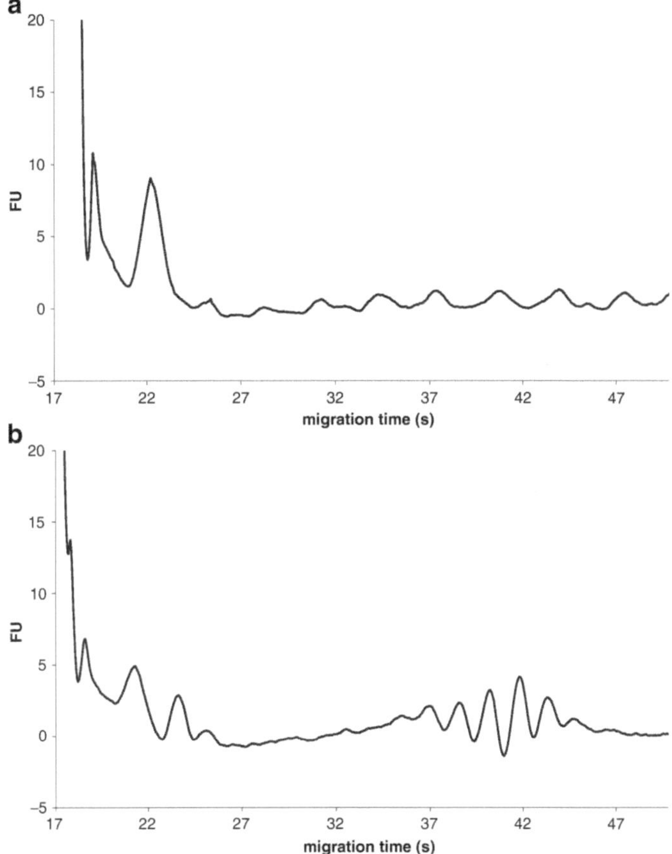

Fig. 3. Microchip electrophoretic profiles of partially purified *S*-LPS samples from *Proteus morganii* 034 (**a**) and *Shigella dysenteriae* 2 (**b**), and of partially purified *R*-LPS samples from *Salmonella enterica* sv. Minnesota R595 whole-cell lysates (**c–d**). The endotoxins are analyzed in the form of endotoxin–dodecyl sulfate–fluorescent dye complexes employing the *Agilent Protein 80 LabChip* kit (**a–c**), or as LPS molecules covalently bound to a fluorescent dye employing the *High Sensitivity Protein 250 LabChip* kit (**d**). The electropherograms do not include the system peak. The number and relative amount of the LPS components in the "wave-like" migration profiles of the *S*-LPSs are unique and characteristic to the respective bacterial strains. The *R*-LPS appears as one single peak (directly after the system peak).

3. Always insert the pipette tip to the bottom of the well when dispensing the liquid. Placing the pipette at the edge of the well may lead to poor results.
4. Unbound fluorescent dye present in the capillary channels is diluted at the detection, since a diluting solution consisting of sieving matrix without the fluorescent dye is introduced via a cross-section just before the detection point. This helps to get a better signal-to-noise ratio.
5. Usually, determination of the molecular masses of protein peaks appearing in the electropherograms can be done with the help of protein molecular mass standards, which should be applied in the ladder well of the microchip. However, proteins have different migration properties than endotoxins,

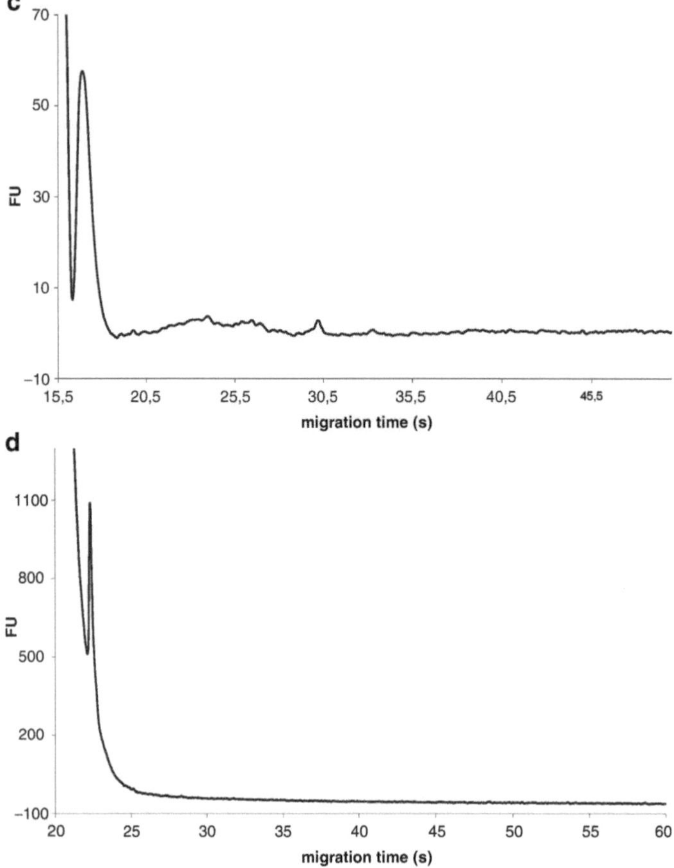

Fig. 3. (continued)

and therefore, the quantitative determination of the molecular masses of LPS peaks obtained by the use of the protein molecular mass standards do not suit the theoretical molecular mass values. At the moment, no proper LPS standards are available.

6. Loaded chips should be used within 5 min as the reagents might evaporate, leading to poor results.
7. The Agilent 2100 bioanalyzer should not be touched during analysis and should never be placed on a vibrating surface.
8. The analysis starts with electrophoretical injection into the capillaries for ca. 15 s (the injected volume is ca. 40 pL) and the run times are ca. 60 s for each sample (collection of data by the software starts at ca. 10 s).
9. The complex of this fluorescent dye and SDS may sometimes appear as several (two or three) peaks (due to unknown reasons). However, this does not disturb the separation of the LPS components. The sensitivity of this method is high, since satisfactory patterns are obtained from 1 mL of bacterial cell cultures, which contain ca. 10^8 cells and less than 30 ng of LPS.

10. The limit of detection of this method is less than 1 ng of LPS.
11. Leaving the chip for a period longer than 1 h in the bioanalyzer may cause contamination of the electrodes.

Acknowledgments

The work was supported by the Grants GVOP-3.2.1-0168, RET 008/2005, and OTKA-NKTH-NI-68863.

References

1. Holst O, Ulmer AJ, Brade H, Flad HD, Rietschel ET (1996) Biochemistry and cell biology of bacterial endotoxins. FEMS Immunol Med Microbiol 16:83–104
2. Magalhaes PO, Lopes AM, Mazzola PG, Rangel-Yagui C, Penna TCV, Pessoa A (2007) Methods of endotoxin removal from biological preparations: a review. J Pharm Pharm Sci 10:388–404
3. Hitchcock PJ, Brown TM (1983) Morphological heterogeneity among Salmonella lipopolysaccharide chemotypes in silver-stained polyacrylamide gels. J Bacteriol 154:269–277
4. Tsai CM, Frasch CE (1982) A sensitive silver stain for detecting lipopolysaccharides in polyacrilamide gels. Anal Biochem 119:115–119
5. Kilár A, Péterfi Z, Csorba E, Kilár F, Kocsis B (2008) Capillary electrophoresis chips for screening of endotoxin chemotypes from whole-cell lysates. J Chromatogr A 1206: 21–25
6. Makszin L, Kilár A, Kocsis B, Kilár F (2009) Fluorescently labeled bacterial endotoxins in microchip electrophoresis. *Electrophoresis – submitted for publication*
7. Bousse L, Mouradian S, Minalla A, Yee H, Williams K, Dubrow R (2001) Protein sizing on a microchip. Anal Chem 73: 1207–1212

Chapter 9

Isolation of Smooth-Type Lipopolysaccharides to Electrophoretic Homogeneity

Elder Pupo

Abstract

Polyacrylamide slab gel electrophoresis in the presence of sodium dodecyl sulfate or sodium deoxycholate (SDS- or DOC-slab-PAGE) is a powerful technique for the separation of smooth(S)-type bacterial lipopolysaccharides (LPS). In order to recover the individual LPS species from the polyacrylamide gel for subsequent analyses, a sensitive, nondestructive reverse staining of slab-PAGE-separated LPS has been developed. The individual reverse-stained LPS bands can be rapidly and efficiently recovered into an aqueous 5% triethylamine solution when they are extruded to produce fine gel microparticles. Based on these principles, an isolation methodology that combines preparative slab-PAGE, reverse staining, extrusion, and passive elution can be used to isolate, to electrophoretic homogeneity, micrograms to hundreds of micrograms of individual LPS species successfully from smooth-type LPS mixtures.

Key words: Lipopolysaccharide, Smooth, Isolation, Electrophoresis, Negative staining, Passive elution

1. Introduction

The fractionation of smooth(S)-type lipopolysaccharides (LPS) into their different molecular species is extremely difficult due to their high structural heterogeneity and their amphipathic character, the latter resulting in the tendency to form molecular aggregates. Polyacrylamide slab gel electrophoresis in the presence of sodium dodecyl sulfate or sodium deoxycholate (SDS- or DOC-slab-PAGE) is the method of choice for the resolution of S-type LPS in a wide molecular mass range. However, slab-PAGE of LPS had not been widely used for micro-preparative purposes, since silver staining – the conventional and most sensitive (1–10 ng/band) method used for visualizing polyacrylamide gel-separated LPS (1) – causes chemical modification and irreversible fixation of

Otto Holst (ed.), *Microbial Toxins: Methods and Protocols*, Methods in Molecular Biology, vol. 739,
DOI 10.1007/978-1-61779-102-4_9, © Springer Science+Business Media, LLC 2011

LPS in the gel, hindering the recovery of intact molecules for subsequent bioanalyses. On the contrary, the fractionation of the slab-PAGE-separated LPS banding pattern by sectioning the unstained gel into two or three regions (2, 3) has proved useful for separating rough-type from smooth-type LPS fractions, but it did not allow a precise fractionation of individual LPS bands. In fact, purified LPS fractions generally showed several LPS species upon re-electrophoresis (2, 3).

To overcome these obstacles, a sensitive, nondestructive reverse staining of slab-PAGE-separated LPS (4) was established. During staining, a white precipitate of zinc–imidazole complex is formed all along the gel surface except on the zones containing LPS that appear as transparent, colorless bands. Negatively stained bands are clearly observable by placing the gel above any dark background. The individual reverse-stained LPS bands can be rapidly and efficiently recovered by passive diffusion in 5% triethylamine when they are extruded to produce fine gel microparticles (5, 6). Based on these principles, an isolation methodology that combines preparative slab-PAGE, reverse staining, extrusion, and passive elution can be used to isolate, to electrophoretic homogeneity, micrograms to hundreds of micrograms of individual species successfully from complex LPS mixtures (7–9).

2. Materials

2.1. Sodium Deoxycholate-Slab-Polyacrylamide Gel Electrophoresis

1. Sample buffer (2×): 10 mL of 0.1 M Tris–HCl buffer, pH 6.8, containing 1% (w/v) DOC, 20% (v/v) glycerol, and 0.005% bromophenol blue. Store at –20°C in 1 mL aliquots.
2. Separating buffer (4×): 1 L of 1.5 M Tris–HCl, pH 8.7. Store at 20–25°C.
3. Stacking buffer (4×): 1 L of 0.5 M Tris–HCl, pH 6.8. Store at 20–25°C.
4. *N,N,N,N'*-Tetramethyl-ethylenediamine (TEMED) and 30% acrylamide/bis solution (29:1, 3.3%). Store at 4°C (see Note 1).
5. Ammonium persulfate: prepare 10 mL of 10% (w/v) solution in water and store at –20°C in 200-μL aliquots for single use.
6. Water-saturated isobutanol. Mix equal volumes (e.g., 5 mL) of water and isobutanol and allow the two phases to separate. Use the upper layer. Store at 20–25°C and in a clear container so that the interface is visible.
7. Running buffer (10×): 1 L of 0.25 M Tris and 1.92 M glycine. Store at 4°C.
8. Sodium deoxycholate: prepare 500 mL of 10% (w/v) solution in water. Store at 20–25°C.

2.2. Reverse Staining of Gel-Separated LPSs

1. Acetonitrile solution (30%): Mix 180 mL acetonitrile and 420 mL water in a graduated cylinder. This solution is prepared just before use.
2. Imidazole (10×) solution: 500 mL of 2 M imidazole. Store at 20–25°C in an opaque container.
3. Zinc sulfate (100×) solution: 500 mL of 1.5 M zinc sulfate. Store at 20–25°C.

2.3. Mobilization of Reverse-Stained LPS

1. Zinc-chelating solution (5×): 1 L of 500 mM EDTA, pH 8. Dissolve the EDTA by adjusting the pH to 8 with sodium hydroxide while stirring. Note that the EDTA will not completely dissolve until the pH is around 8. Add sodium hydroxide slowly to avoid exceeding this pH value. Store at 20–25°C.

2.4. Elution of LPSs

1. Gel extruder (1 mL capacity): To prepare this device, the following materials are needed: a 1-mL polypropylene syringe, two pieces of metal sieves (2 × 2 cm in size) with holes of 125 and 32 μm in diameter (Carl Schroeter, Hamburg, Germany), scissors, a surgical blade, and a pencil. Take the plunger out of the syringe. Then, with the surgical blade, cut the tip of the syringe at its base. Place the metal sieves on a flat surface and draw on each of them, with a pencil, a circle with a diameter 0.5 mm greater than the internal diameter of the syringe. Cut out the two circles with the scissors. Insert the disk of 32-μm metal sieve at the bottom of the syringe with the aid of the plunger. The second disk of 125-μm metal sieve is placed over the 32-μm disk and firmly adjusted by pressing them with the plunger. The gel extruder includes also the plunger and this latter is left inside the barrel of the syringe.
2. Gel extruder (5 mL capacity): The following materials are required: a 5-mL polypropylene syringe, one piece of metal sieve (2 × 2 cm in size) with holes of 125 μm in diameter (Carl Schroeter, Hamburg, Germany), scissors, a surgical blade, and a pencil. Take the plunger out of the syringe. Then, with the surgical blade, cut the tip of the syringe at its base. Perforate the bottom of the syringe by cutting out three separate triangular sections. The parts of the bottom of the syringe that are left between the perforations should remain linked together since they will serve as a support for the metal sieve. Place the metal sieve on a flat surface and draw on it, with a pencil, a circle with a diameter 0.5 mm greater than the internal diameter of the syringe. Cut out the circle with the scissors. Insert the disk of the 125-μm metal sieve at the bottom of the syringe with the aid of the plunger. Leave the plunger inside the barrel of the syringe (see Note 2).

3. Elution solution: Just before use, prepare 100 mL of a 5% triethylamine solution in water (see Note 3).
4. Neutralization solution: Acetic acid (glacial) 100%. Handle this reagent in the fume hood.

2.5. Silver Staining of LPS

1. Acetonitrile solution (30%): Combine 180 mL acetonitrile and 420 mL water.
2. Fixation and oxidation solution: 200 mL of a solution containing 0.7% sodium metaperiodate, 40% (v/v) ethanol, and 5% (v/v) acetic acid.
3. Staining solution (see Note 4): Add 2 mL of 25% ammonia solution to 28 mL of 0.1 M sodium hydroxide. Dissolve 1 g of silver nitrate in 5 mL water and add this to the solution while it is being stirred. A transient brown precipitate will form and disappear within seconds. Add 115 mL water to make a total of 150 mL (see Note 5).
4. Developer solution: Dissolve 10 mg citric acid in 200 mL water and add to this 0.1 mL of 37% (v/v) formaldehyde (see Note 6).
5. Stop solution: 200 mL of 10% (v/v) acetic acid.

3. Methods

The method for isolation of LPS species consists of five main steps: preparative DOC-slab-PAGE, reverse staining of gel-separated LPS, LPS mobilization, LPS elution, and centrifugal ultrafiltration. The whole procedure, from the application of the initial LPS mixture to preparative DOC-PAGE to the analysis of the isolated LPS fractions by re-electrophoresis, can be completed in 3 days. The different steps of the procedure are to be divided in matters of time as follows: day 1, preparative DOC-PAGE and reverse staining of gel-separated LPS; day 2, LPS mobilization, LPS elution, and centrifugal ultrafiltration; day 3, analytical DOC-PAGE and silver staining.

The procedure is exemplified here by using the smooth-type LPS mixture from *Escherichia coli* K-235 (Sigma, St. Louis, MO). Glycine DOC-PAGE is utilized for the separation of LPS, since this system has previously shown to dissociate S-type LPS molecules more efficiently compared to glycine SDS-PAGE (10). Another advantage of DOC-PAGE is that boiling of the LPS sample prior to electrophoresis is not necessary which is beneficial for preserving the chemical integrity of LPS. The separation conditions of LPS by preparative DOC-PAGE, such as the quantity of LPS applied on the preparative gel and the acrylamide concentration of the separating gel should be selected to produce an adequate resolution of the LPS species of interest. High-purity, freshly produced deionized water with a resistivity of 18.2 MΩ cm is used for the preparation of all solutions.

3.1. Preparative DOC-Slab-PAGE

1. These directions consider the use of a PROTEAN II xi Cell system (Bio-Rad, Hercules, CA), glass plates of dimensions 16 × 16 cm, and 1-mm-thick spacers. For preparative electrophoresis, a special preparative comb is used that contains a single wide tooth. The glass plates should be washed with detergent, thoroughly rinsed with water, and wiped with clean paper such as Kimwipes wetted with ethanol. Rinsed gloves should be worn while preparing and handling the gels to prevent fingerprints.
2. Assemble the glass plates and spacers in the gel casting stand as described by the manufacturer.
3. Prepare a 17% separating gel solution by mixing 7.5 mL of 4× separating buffer, with 17 mL of 30% acrylamide/bis solution, 1.5 mL of 10% sodium deoxycholate, 3.85 mL water, 150 μL of 10% ammonium persulfate, and 15 μL TEMED (see Note 7). Carefully pour the gel mixture between the glass plates to the desired volume, and remember to leave space for the stacking gel. Layer a small amount of water-saturated isobutanol onto the top of the gel. Allow polymerization of the gel to proceed for 30 min.
4. Pour off the isobutanol and rinse the top of the gel twice with water.
5. Prepare a 4% stacking gel by mixing 2.5 mL of 4× stacking buffer, with 1.3 mL of 30% acrylamide/bis solution, 0.5 mL of 10% sodium deoxycholate, 5.7 mL water, 50 μL of 10% ammonium persulfate, and 10 μL TEMED. Pour the stacking gel mixture on top of the separating gel and insert the comb. Allow the gel to polymerize completely (about 30 min) before removing the comb.
6. Prepare 2 L of the running buffer by mixing 200 mL of the 10× running buffer with 1.8 L water in a graduated cylinder.
7. Carefully remove the comb and use a 3-mL syringe fitted with a 22-gauge needle to wash the wells with running buffer.
8. Prepare the LPS sample for preparative electrophoresis by mixing a suspension of 500 μg smooth-type LPS in water with the same volume of 2× sample buffer.
9. Add the running buffer to the upper and lower chambers of the gel apparatus and carefully layer the LPS sample into the wide gel well.
10. Complete the assembly of the electrophoresis unit according to the manufacturer's specifications and connect the electrodes

to a power supply. Perform the electrophoresis run at 15 mA at 20–25°C until the dye reaches approximately 1 cm from the bottom of the gel.

3.2. Reverse Staining of Gel-Separated LPSs

1. After electrophoresis, the gel-separated LPS are detected as transparent, colorless bands contrasting against a white gel background by staining with zinc–imidazole. A transparent plastic or glass staining dish of appropriate size for the gel should be cleaned with detergent and thoroughly rinsed with water. Wear gloves rinsed with water to handle the gel. All washing and incubation steps are carried out under gentle shaking at 20–25°C (see Note 8).
2. Disassemble the electrophoresis unit. Place the gel sandwich on a flat surface. Remove the spacers and the top glass plate. The stacking gel is cut out and discarded. Place the separating gel in a staining dish.
3. Wash the separating gel three times for 10 min with 30% acetonitrile to substantially remove DOC and other electrophoresis-associated chemicals.
4. Wash the gel three times for 10 min with water.
5. Prepare 200 mL of 1× zinc sulfate solution by adding 2 mL of 100× zinc sulfate solution to 198 mL of water.
6. Incubate the gel in 1× zinc sulfate solution for 15 min (see Note 9).
7. Prepare 200 mL of 1× imidazole solution by adding 20 mL of 10× imidazole solution to 180 mL of water.
8. Discard the zinc sulfate solution and wash the gel for not more than 3–5 s with fresh water.
9. Soak the gel in 1× imidazole solution for 1–3 min. A white precipitate of zinc–imidazole forms all along the gel surface except on the zones containing LPS, which appear as transparent, colorless bands. The negatively stained bands are clearly observable by placing the gel above any dark background (see Note 10).
10. Wash the gel three times for 10 min with water (see Note 11).
11. Store the stained gel in water at 4°C until use (see Note 12). An example of the staining pattern of LPS produced on the preparative gel is shown in Fig. 1.

3.3. Mobilization of Reverse-Stained LPSs

1. Following detection, the LPS bands of interest are excised. To do this, place the gel on a clean glass plate and hold it a few centimeters above a dark surface to visualize the staining pattern. Carefully cut out each LPS band along its contour with the aid of a surgical blade. Leave always a small section of the gel between adjacent LPS bands to prevent contamination.

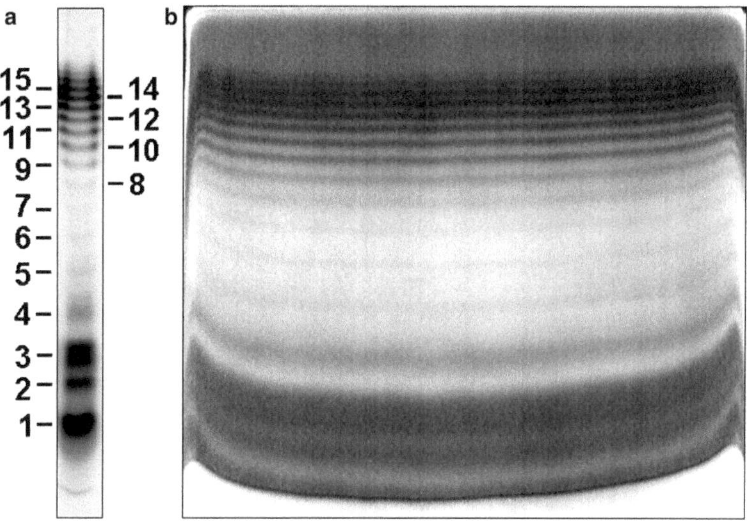

Fig. 1. DOC-PAGE (17%) separation of LPS species from *E. coli* K-235. (**a**) Analytical separation of 6.25 μg of LPS. Fifteen bands are identified and numbered in order of increasing LPS molecular mass. The gel was stained with silver. (**b**) Preparative separation of 500 μg of LPS. The gel-separated LPS were reverse stained by using zinc and imidazole. (Reproduced from ref. (9) with permission from Wiley-VCH).

2. Transfer the gel slices to separate Corning® 50-mL clear polypropylene centrifuge tubes.
3. To chelate zinc ions, place the tubes on an overhead mixer and incubate the LPS bands under agitation twice for 10 min in 45 mL of the 1× zinc-chelating solution.
4. Then, wash the LPS bands three times for 10 min under agitation with 45 ml of water to remove the chelating solution thoroughly.

3.4. Elution of LPSs

1. Finely cut the gel slices with a surgical blade so that they fit into the barrel of the 1-mL gel extruder. Alternatively, the gel slices can be converted into smaller pieces by using a 10-mL polypropylene syringe without needle. To do this, transfer the gel slices into a 10-mL syringe and press them with the plunger against the bottom of the syringe. Smaller gel pieces will come out of the tip of the syringe.
2. Transfer the gel pieces into the 1-mL gel extruder. If a 10-mL syringe is used in the previous step, this can be done simply by introducing the tip of the 10-mL syringe into the barrel of the 1-mL extruder. Extrude the gel pieces through the metal sieves by pressing the plunger. Collect the resulting gel microparticles in 50-mL polypropylene tubes (see Note 13).
3. To passively elute LPS molecules, add a volume of 5% (v/v) triethylamine equal to three times that of the extruded gel.

Then, place the tubes on an overhead mixer and agitate the gel slurries for 5 min at 20–25°C (see Note 14).

4. Immediately afterward, in the fume hood, pour the gel slurry in a small beaker and agitate the mixture with a magnetic stirrer. While stirring, add a volume of the neutralization solution (acetic acid 100%) equal to 1/52 that of the 5% triethylamine solution initially added to the extruded gel. Monitor the pH with a pH-meter and slowly add a few more drops of the neutralization solution until a pH value of 7 is reached (see Note 15). Stir the gel slurry in the neutralized elution solution for 5 min at 20–25°C.
5. Centrifuge the tubes for 5 min at 1,200 × *g* and gently collect with a pipette the overlaying solution which contains the eluted LPS. Take good care not to remove any gel microparticle.
6. Add to the gel microparticles a volume of water equal to that of the overlaying solution collected in the previous step. Stir the gel slurry for 5 min at 20–25°C.
7. Centrifuge the tubes for 5 min at 1,200 × *g* and carefully collect with a pipette the overlaying solution which contains the eluted LPS. Take good care not to remove any gel microparticle.
8. Pool the samples collected from the two LPS elution steps.
9. To remove any remaining gel microparticle, the LPS sample is filtered through a 5-μm pore-size syringe filter (Whatman).

3.5. Centrifugal Ultrafiltration

1. To remove the triethylammonium acetate salt from the LPS, perform extensive diafiltration of the sample against water at 4°C using a 15-mL capacity Centriprep® centrifugal filter unit with a 10-kDa cut-off membrane (Millipore). First, concentrate the sample to 1 mL by centrifuging the filter unit at 2,000 × *g*. The triethylammonium acetate salt passes through the membrane while the LPS species are kept in the retentate. Then add 14 mL water to the retentate and concentrate again the sample to 1 mL under centrifugation. Repeat this ultrafiltration sequence four times.
2. Further concentrate the LPS sample by utilizing a Microcon® centrifugal filter unit (10 kDa molecular mass cut-off, Millipore). Transfer the LPS sample into the filter unit and centrifuge at 12,000 × *g* at 4°C until 20–50 μL of sample is reached. The concentrated LPS sample is recovered with a micropipette, transferred to a clean vial, and stored at –20°C until use.

3.6. Analytical DOC-PAGE

1. The separation of the isolated LPS fractions by gel electrophoresis is performed just as already described for preparative DOC-slab-PAGE, taking into account the following modifications: (1) volumes of the 30% acrylamide/bis solution and water of 15 and 5.85 mL, respectively, are used for the

preparation of the separating gel solution (15%); (2) a 25-teeth comb is used for casting the gel wells; and (3) aliquots of 5–15 μL of the isolated LPS fractions and one aliquot of the initial unfractionated mixture containing 12.5 μg of LPS, to be used as a reference, are loaded onto the gel wells.

3.7. Silver Staining of LPS Separated by DOC-PAGE

1. The gel-separated LPS are visualized by silver staining. A glass staining tray of appropriate size for the gel should be cleaned with 50% nitric acid after cleaning with detergent, and thoroughly rinsed with high-quality water. To avoid fingerprints, wear gloves rinsed with water to handle the gel. All washing and incubation steps are carried out under gentle shaking at 20–25°C. Only high-purity deionized or distilled water should be used.
2. Disassemble the electrophoresis unit. Place the gel sandwich on a flat surface. Remove the spacers and the top glass plate. The stacking gel is cut out and discarded. Place the separating gel in a staining tray.
3. To remove excess DOC and avoid interference in the technique, wash the gel three times for 10 min with 200 mL of 30% acetonitrile. Then, wash the gel three times for 10 min with 200 mL of water.
4. Incubate the gel in the fixation and oxidation solution under agitation for 30 min.
5. Wash the gel three times for 10 min with 200 mL of water.
6. Incubate the gel in the staining solution under agitation for 25 min. Next, discard the staining solution into an appropriate liquid waste container.
7. Wash the gel three times for 10 min with 100 mL of water.
8. Incubate the gel in the developer solution and wait for a few minutes until adequate stain intensity is reached.
9. Discard the developer solution immediately and incubate the gel in the stop solution under agitation for 10 min.
10. Wash the gel three times for 10 min with 200 mL of water. An example of the obtained gel image, as recorded with a scanner, is shown in Fig. 2. This demonstrates the electrophoretic homogeneity of the LPS fractions isolated by the present methodology.

4. Notes

1. Caution: non-polymerized acrylamide is a neurotoxin. Care should be taken to prevent exposure. Avoid inhalation when preparing the acrylamide solution from the powdered reagent;

Fig. 2. DOC-PAGE (15%) analysis of LPS fractions from *E. coli* K-235. Fifteen LPS bands on the preparative gel in Fig. 1b were separately isolated and re-electrophoresed in an analytical (15%) DOC-PAGE. Lanes (C): 12.5 μg of the unfractionated LPS; (1)–(15): purified LPS bands. The gel was stained with silver. The recovery of LPS bands from six preparative gels (3 mg unfractionated LPS) was as follows: for the major LPS bands of low and high molecular mass, quantities in the range from 82 to 411 μg and from 20 to 84 μg, respectively, were obtained. For the minor LPS bands of intermediate molecular mass, quantities were recovered, ranging from 0.28 to 18.9 μg. (Reproduced from ref. (9) with permission from Wiley-VCH).

it is advisable to purchase the premixed solution (500 mL, Bio-Rad, Hercules, CA). Wear gloves at all times.

2. The inclusion of a second disk of 32-μm metal sieve in the 5-mL extruder is not convenient, since this notably increases the pressure that has to be exerted on the plunger for extrusion and makes manipulation of the extruder extremely difficult.
3. This solution has a strong ammonia-like odor and is toxic by inhalation. It should be handled in the fume hood and stored in sealed containers. Avoid exposure through the skin or the eyes. The triethylamine stock solution is stored at 4°C in an opaque flask. Verify that this solution does not have a deep yellow color (indicative of a significant deterioration).
4. This staining reagent is toxic and corrosive and should be prepared in the fume hood.
5. Prevent deposition of this chemical on the skin as it may produce, depending on the extent of exposure, from mild skin stains to more severe burns.
6. Avoid any kind of exposure to formaldehyde since this is a toxic and carcinogenic substance.
7. For preparing a different gel concentration (denoted as X), add X mL of 30% acrylamide/bis solution and (20.835-X) mL water. The quantities of the rest of the ingredients remain unchanged.

8. The quality of the water is critical for the proper development and stability of zinc–imidazole reverse staining. Only freshly produced high-purity deionized water with a resistivity of 18.2 MΩ cm is used in all steps. Changes in the pH of water, specially acidification, may lead to variable qualities in the gel image and to the loss of the staining pattern after incubation for several hours.
9. Care should be taken to prepare the zinc sulfate solution at the indicated concentration (15 mM). Notice that incubating the gel at zinc sulfate concentrations of 5 mM or less produces an insufficient development of the staining and a low image contrast, which hampers sensitivity. Conversely, at a concentration of 100 mM or higher, the sensitivity of zinc decreases due to gel overstaining. Zinc concentrations ranging from 10 to 20 mM strike a satisfactory balance between sensitivity and intensity of background staining.
10. The sensitivity of this stain (1–5 ng/band) is similar to that of silver. Clear digital images of the reverse-stained gel can be easily stored by scanning the gel in a darkened room while keeping the scanner lid open. Alternatively, a photographic recording can be performed by placing the gel on a glass plate some centimeters above a black surface. Illuminate evenly the gel with fluorescent lights. The photograph is taken from above.
11. Washing with water is required to remove excess imidazole in the gel. Longer incubation of the gel in the imidazole solution may cause overstaining of the background with loss of sensitivity for minor components.
12. Stained gels can be stored in water at 4°C without fading for at least 1 week.
13. When isolating LPS bands from several, simultaneously run preparative gels, the total volume of gel may be several times greater than the capacity of the 1-mL extruder. In this case, replace the 1-mL extruder with the 5-mL capacity gel extruder.
14. In our experience, the incubation of LPS for 5 min in 5% triethylamine at 25°C does not produce any significant chemical modification of LPS. Nevertheless, the chemical integrity of LPS, specially the content of labile *O*-linked fatty acids, should be verified with appropriate (e.g., mass spectrometry of intact LPS) analytical techniques.
15. The 5% triethylamine solution has a pH of 12.4. Under this condition, *O*-acetyl groups of LPS are released in a time-dependent manner. Therefore, after 5-min incubation, the elution solution should be neutralized to prevent any significant chemical modification of the LPS.

Acknowledgments

The author is a research fellow of the Alexander von Humboldt Foundation (Germany). The contribution of Prof. Dr. E. Hardy (University of Havana, Cuba) to the development of the methodology described here is specially acknowledged.

References

1. Tsai CM, Frasch CE (1982) A sensitive silver stain for detecting lipopolysaccharides in polyacrylamide gels. Anal Biochem 119:115–119
2. Ohta M, Rothmann J, Kovats E, Pham PH, Nowotny A (1985) Biological activities of lipopolysaccharide fractionated by preparative acrylamide gel electrophoresis. Microbiol Immunol 29:1–12
3. Komuro T, Yomota C, Kimura T, Galanos C (1989) Comparison of R- and S-form lipopolysaccharides fractionated from *Escherichia coli* UKT-B lipopolysaccharide in pyrogen and Limulus tests. FEMS Microbiol Lett 51:79–83
4. Hardy E, Pupo E, Castellanos-Serra L, Reyes J, Fernandez-Patron C (1997) Sensitive reverse staining of bacterial lipopolysaccharides on polyacrylamide gels by using zinc and imidazole salts. Anal Biochem 244:28–32
5. Hardy E, Pupo E, Santana H, Guerra M, Castellanos-Serra LR (1998) Elution of lipopolysaccharides from polyacrylamide gels. Anal Biochem 259:162–165
6. Pupo E, Lopez CM, Alonso M, Hardy E (2000) High-efficiency passive elution of bacterial lipopolysaccharides from polyacrylamide gels. Electrophoresis 21:526–530
7. Pupo E, Phillips NJ, Gibson BW, Apicella MA, Hardy E (2004) Matrix-assisted laser desorption/ionization-time of flight-mass spectrometry of lipopolysaccharide species separated by slab-polyacrylamide gel electrophoresis: high-resolution separation and molecular weight determination of lipooligosaccharides from *Vibrio fischeri* strain HMK. Electrophoresis 25:2156–2164
8. Gulin S, Pupo E, Schweda EK, Hardy E (2003) Linking mass spectrometry and slab-polyacrylamide gel electrophoresis by passive elution of lipopolysaccharides from reverse-stained gels: analysis of gel-purified lipopolysaccharides from *Haemophilus influenzae* strain Rd. Anal Chem 75:4918–4924
9. Pupo E, Hardy E (2007) Isolation of smooth-type lipopolysaccharides to electrophoretic homogeneity. Electrophoresis 28:2351–2357
10. Komuro T, Galanos C (1988) Analysis of *Salmonella* lipopolysaccharides by sodium deoxycholate-polyacrylamide gel electrophoresis. J Chromatogr 450:381–387

Chapter 10

A Method for Unobtrusive Labeling of Lipopolysaccharides with Quantum Dots

Carlos Morales-Betanzos, Maria Gonzalez-Moa, and Sergei A. Svarovsky

Abstract

Bacterial endotoxins or lipopolysaccharides (LPS) are among the most potent activators of innate immune system, yet mechanisms of their action and, in particular, the role of the glycans remain elusive. Efficient noninvasive labeling strategies are necessary for studying interactions of LPS glycans with biological systems. Here, we describe a new method for labeling LPS and other lipoglycans with luminescent quantum dots (QDots). The labeling is achieved by the partitioning of hydrophobic quantum dots into the core of various LPS aggregates without disturbing the native LPS structure. The biofunctionality of the LPS–QDot conjugates is demonstrated by labeling of mouse monocytes. This simple method will find broad applicability in studies concerned with visualization of LPS biodistribution and identification of LPS-binding agents.

Key words: Endotoxin, Lipopolysaccharide, Labeling, Monocytes, Quantum dots, Fluorescence imaging

1. Introduction

Bacterial lipopolysaccharides (LPS), also known as endotoxins, are the major constituents of the outer surface of Gram-negative bacteria (1). They occupy up to 90% of the bacterial cell surface and are responsible for septic shock, which kills nearly 200,000 critically ill patients each year in the USA alone (2). Not surprisingly, there is a great deal of interest in understanding mechanisms of LPS action for the developing of antisepsis drugs. The development of such agents depends on the availability of efficient labeling strategies for LPS molecules (3). Ideally, such labeling should not be disruptive to the LPS functionality.

Otto Holst (ed.), *Microbial Toxins: Methods and Protocols*, Methods in Molecular Biology, vol. 739, DOI 10.1007/978-1-61779-102-4_10, © Springer Science+Business Media, LLC 2011

LPS are complex, negatively charged lipoglycans composed of three distinct regions: (a) a fatty acid region called lipid A; (b) a core region oligosaccharide composed of up to 15 monosaccharides; and (c) a highly variable O-antigenic polysaccharide responsible for much of the bacterial pathogenicity and immunospecificity. Most labeling strategies rely on chemical modification of LPS molecules with organic dyes and normally require complex manipulations and purification steps (4, 5) due to the aggregative tendencies of LPS molecules (6). The chemical modification is not site-specific and depends on the availability of reactive groups that are not always accessible or available in LPS (1). If such groups are not present, they are chemically introduced by oxidation of the O-antigenic glycans (5, 7). By introducing additional moieties to the LPS molecule, these methods perturb its physical properties and biomolecular recognition events (8), making such probes unlikely candidates for elucidating the roles of glycan interactions.

Nanometer-sized crystals of semiconductors known as quantum dots (QDots) have recently emerged as useful luminescent labeling agents (9). These nanoprobes have significant benefits over organic dyes including long-term photostability, high luminescent intensity, and multiple colors with single-wavelength excitation that opens up possibilities for multiplex detection. Coating of hydrophobic QDots with phospholipids (10) and synthetic amphiphilic polymers has been described (11). Both methods rely on phase transfer of hydrophobic QDots from an organic solvent to an aqueous solution of amphiphilic molecules. Here, we describe an application of hydrophobic QDots to non-covalent labeling of LPS and its derivatives. We show that this method may be broadly applicable to other lipoglycans as well. This method takes advantage of the universal amphiphilic nature of lipoglycans and does not introduce any chemical modifications to the LPS structure, making it ideally suitable for studying glycan interactions.

2. Materials

2.1. Labeling of LPS with QDots

1. Organic Quantum Dots (QDot® 605 ITK™, Invitrogen, Inc) (see Note 1).
2. Smooth-type LPS from *Escherichia coli* serotypes O111:B4 and O55:B5 (Sigma Inc.) and *Pseudomonas aeruginosa* (Sigma Inc.), as well as lipid A (Avanti Polar Lipids, Inc.), Kdo_2-lipid A (Avanti Polar Lipids, Inc.), and lipoteichoic acid (LTA) from *Bacillus subtilis* (Sigma, Inc.) (see Note 2) (Fig. 1).
3. Chloroform and methanol.

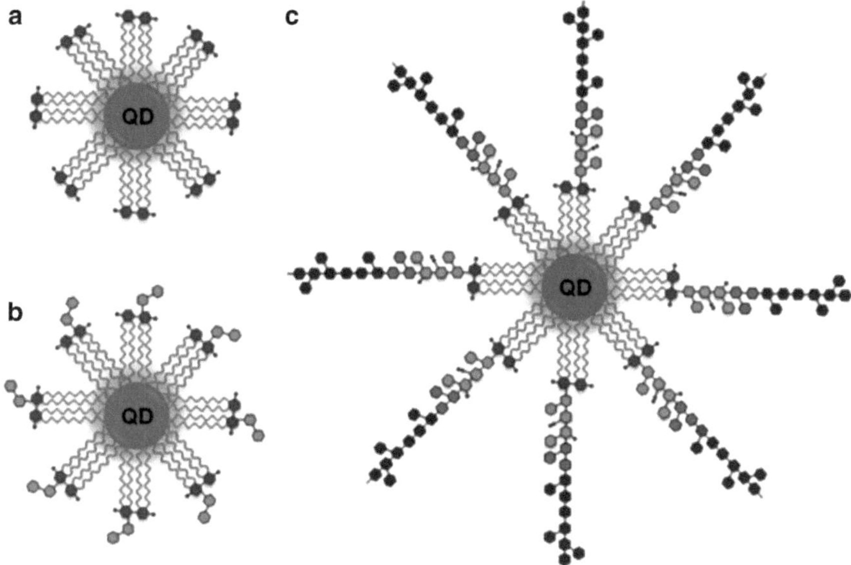

Fig. 1. Schematic of the lipoglycans–QDot micelles formed with (**a**) lipid A, (**b**) Kdo_2-lipid A, and (**c**) LPS.

4. A saturated solution of tetramethylammonium hydroxide pentahydrate ($Me_4NOH+5H_2O$).
5. Deionized water (18 mΩ) was obtained from a Millipore ultrapure water filtration unit.
6. Zeba desalting spin columns (Pierce Inc.).
7. Speed Vacuum concentrator (Thermo Scientific, Inc.).

2.2. Cell Culture

1. Mouse monocytes (ATCC RAW 264.7, American Type Culture Collection, Manassas, VA).
2. Dulbecco's modified Eagle medium (DMEM, GIBCO, Grand Island, NY) supplemented with 10% fetal calf serum (FCS).
3. Dulbecco's phosphate-buffered saline 1× (DPBS, GIBCO, Grand Island, NY).
4. Nonenzymatic cell dissociation agent (Cellstripper™ Mediatech Inc.).
5. Hank's balanced salt solution (HBSS, GIBCO, Grand Island, NY).
6. Fixing solution – paraformaldehyde 1% in PBS (Invitrogen Inc.).

2.3. Flow Cytometry

1. Flow cytometer (FACS Caliber, BD Biosciences Inc) (see Note 3).
2. 5 mL polystyrene round-bottom tube 12 × 75 mm style.

2.4. Fluorescence Microscope

1. Fluorescence microscope (for example, BX51, Olympus America Inc) (see Note 4).
2. Poly(L-lysine)-coated glass slides (see Note 5).

2.5. Dynamic Light Scattering

1. Zetasizer NanoZS (Malvern Instruments).
2. 100 μL disposable plastic cuvettes (Malvern Instruments, model ZEN0117).

2.6. Transmission Electron Microscopy

1. Electron microscope (Philips CM12S) operated at an accelerating voltage of 80 kV.
2. Carbon-formvar mesh grids.
3. Digital camera (e.g., Gatan model 791).

3. Methods

The described method for LPS labeling is compatible with several types of LPS varying in the glycan component, as well as for other lipoglycans with a similar amphipatic behavior. The resulting bioluminescent probes in the form of micelles are better evaluated by dynamic light scattering (DLS) and transmission electron microscopy (TEM). Although the protocol as described was designed for complete LPS, we have also used it for labeling monophosphoryl lipid A, which is the amphipatic component of LPS responsible for its toxicity, Kdo_2-lipid A, and LTA. Although lipid A does not contain the glycan elements present in complete LPSs, it also forms micelles in solution, which prove to be necessary for our labeling method.

3.1. Labeling of LPS in Suspension

1. Evaporate the solution of the Organic QDots (Invitrogen, Inc) until dry using the Speed Vac at room temperature (23°C). Resuspend the solid residue in equal amount of chloroform. In our knowledge, the QDs suspended in chloroform are stable for several months without any visible sign of flocculation.
2. Take a 100 μL aliquot of the chloroform solution and dilute it to 500 μL with chloroform. Prepare 100 μL of a 10 mg/mL aqueous solution of the corresponding lipoglycan (i.e., *E. coli* 0111:B4, *E. coli* 055:B5, *P. aeruginosa* LPS, or lipid A).
3. Mix together the QDots in chloroform and the lipoglycan solution. At this point, a very distinctive two-layer water–chloroform system will be formed. Add methanol dropwise with occasional vortexing until complete mixing of the two phases (ca. 400 μL of methanol). The homogeneous mixture chloroform/methanol/water will have an approximate ratio of 5:4:1 (see Note 6).
4. Evaporate the homogeneous mixture until dry and resuspend the solid residue in 100 μL of ddH_2O. Add

dropwise tetramethylammonium hydroxide pentahydrate ($Me_4NOH.5H_2O$) solution until pH 11 (ca. 25 μL). The latter basification step is critical as it allows transfer of the QDots into the aqueous phase. No transfer occurs when no lipoglycans are present in the mixture.

5. Then, sonicate the solution for 30 min. Pass the solution through two consecutive Zeba columns to remove salts and excess of free LPS (see Note 7).

3.2. DLS Analysis of Lipoglycan–QDot Micelles

1. Place 100 μL of the labeled LPS solution in a plastic disposable cuvette.
2. Load the cuvette in the Zetasizer, close the lid, and run the size measurement of the lipoglycans–QDot micelles (see Note 8) (Fig. 2).

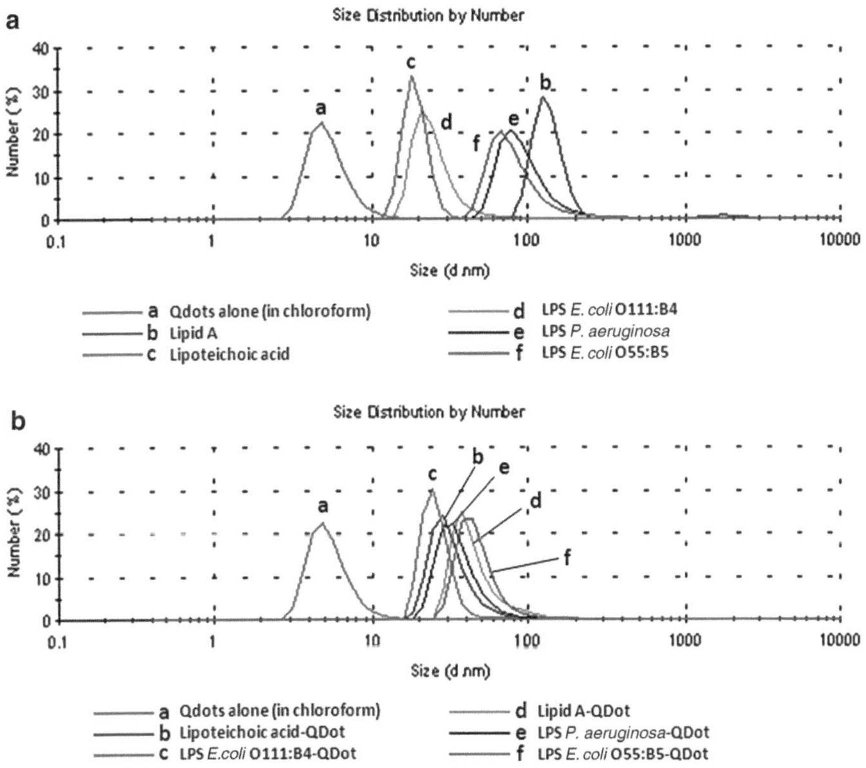

Fig. 2. Size distribution by number of particles obtained from the DLS analysis of the LPS-QDot nanoprobes. The *x*-axis shows the particles diameter in nm, and the *y*-axis shows the percentage of particles of each specific diameter. (**a**) Isolated Qdots and lipoglycans (without Qdots) showing various sizes due to unilamellar and multilamellar arrangements in aqueous solution. (**b**) Isolated QDots and QDots complexed with LPS from *P. aeruginosa* 10, *E. coli* 0111:B4, *E. coli* 055:B5, and lipid A are included. After mixing with QDots and adding base, the size of the newly formed QDots–LPS aggregates is around 50 nm. The basification process makes the LPS monomeric, favoring the access of the QDots to the lipid part of the LPS. Under UV light, the QDot particles are seen in the aqueous solution, which is only possible if the particles are taken up by the amphiphilic LPS. No luminescence is observed in solution prior to adding the base. The sonication of the mixture of QDots–LPS for 30 min makes the aggregate more compact as the diameter is reduced to 38 nm and the peak distribution becomes narrower, indicating increasing homogeneity of the QDot–LPS conjugates. The excess of non-solubilized QDots can be observed under UV light on the walls of the flask.

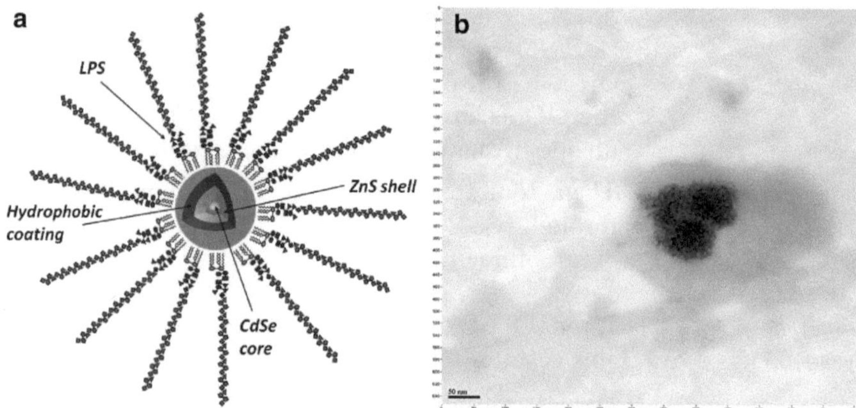

Fig. 3. (**a**) Schematic of LPS–QDot micelles and (**b**) TEM image of the LPS–QDot micelles synthesized using this protocol.

3.3. TEM Analysis of Lipoglycan–QDot Micelles

1. 2 μl of the labeled LPS solution was applied to a glow-discharged mesh copper grid coated with carbon-formvar.
2. The sample was allowed to adhere for 1 min and excess liquid was removed. Contrasting stain was not required due to the inherent electron density of the QDots (Fig. 3).
3. The sample was analyzed at an accelerating voltage of 80 kV in the TEM, and images were recorded using magnification-calibrated digital acquisition.

3.4. QDots–LPS Labeling of Monocytes

1. Culture mouse monocytes until confluence in DMEM supplemented with 10% FCS at 37°C in a 5% CO_2 atmosphere and 95% humidity.
2. Detach the monocytes from the flask using Cellstripper™ and wash twice with 1× PBS. Resuspend the pellet in HBSS 1×. Count the monocytes in the solution and separate them in aliquots of 1 × 106 CFU in 5-mL polypropylene round-bottom tubes. Add LPS pre-labeled with QDots to a final concentration of 100 μg LPS/mL for a final volume of 300 μL. Incubate for 30 min at 37°C.
3. After incubation, wash the cells twice with PBS and fix them in 1% paraformaldehyde solution; wash two more times with PBS to remove the excess fixing solution (Fig. 4).

3.5. Flow Cytometry

1. Place an aliquot of 1×10^6 CFU pre-labeled and fixed in a 5-mL polypropylene round-bottom tube. Place the tube in the flow cytometer's sample injection port.
2. Set the appropriate beam laser for your sample (see Note 4). Set the appropriate filter for the emission of your QDots (see Note 3).
3. Adjust the FSC and SSC detector levels to position the population properly on the graph.

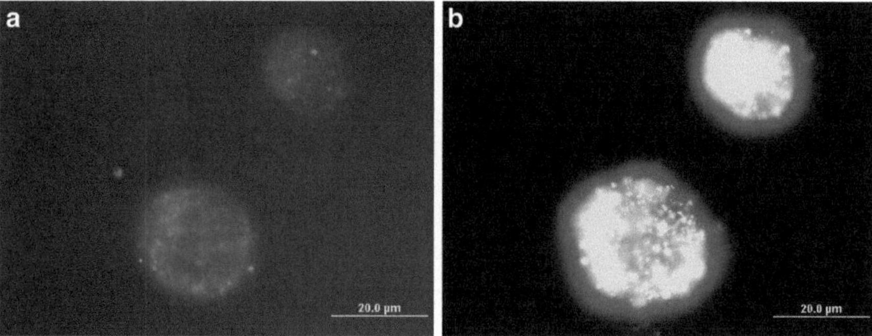

Fig. 4. Fluorescence microscope image of mouse monocytes labeled with QDot–LPS. (**a**) Image taken using a filter with range 480–600 nm. (**b**) Image taken using a filter with a wavelength range of 539–650. QDot 605 nm labeling is only observed in (**b**).

4. Adjust the threshold to eliminate low-level signals.
5. Use the gate tool to draw a region around the cell population of interest.
6. Adjust the fluorescence detector voltage for the channel (color) of interest.
7. Acquire data.

3.6. Fluorescence Microscopy

1. Place 60 μL of the fixed cells solution on a poly-L-lysine-coated glass slide and set a cover slide on top. Allow the cells to attach for 30 minutes at room temperature (23°C).
2. Use the fluorescence microscope to visualize the cells: Set the laser beam and filter in a short visible wavelength, such as blue or green, and view the cells using the corresponding filter for the emission of your QDots (see Note 4). Adjust the exposure time to maximize the signal above the background and acquire images.

4. Notes

1. We used QDot® 605 ITK™ organic QDots from Invitrogen with a peak emission wavelength of 605 nm. These QDots have a lipophilic coating required for this protocol. Alternatively, other organic QDots can be used with a different emission wavelength (Invitrogen offers organic QDots with seven different emission wavelengths: 545, 565, 585, 605, 655, 705, 800 nm)
2. Note that LPS, lipid A, Kdo_2-lipid A, and LTA are pyrogens that may cause fever. It may be harmful if inhaled, ingested, or absorbed through skin. Good laboratory practice should

be employed. Wear a lab coat, gloves, safety glasses and a respirator mask. LPSs from numerous Gram-negative bacteria are available from a large number of vendors. In the method described here, LPS from three different bacteria was used. Additionally, other lipoglycans with a similar amphipatic behavior can be used with our method. For example, monophosphoryl lipid A (PHAD, Avanti Inc.) was used to synthesize labeled micelles with similar properties to the LPSs.

3. We used a Flow Cytometer (FACS Caliber, BD Biosciences Inc) equipped with a 488 nm Argon laser. The instrument has three band-pass emission filters. FL1: 530 ± 30 nm (green), FL2: 585 ± 42 nm (yellow–orange), and FL3 > 650 nm (red). Since the emission of the QDots we used is 605 nm, FL2 was used.
4. We used a fluorescence microscope (BX51, Olympus America Inc) equipped with a 100-W mercury lamp, two separate filter wheels for fluorescence and excitation emission, and an Olympus DP70 12.5 megapixel cooled digital camera (Olympus America Inc.). The samples labeled with QDot® 605 ITK™ were exited using a 488 nm beam, and emission was registered through an orange filter. (QDots are characterized by a broad excitation band that precedes the emission peak and are better excited using UV light. However, for cell samples, it is recommended to use short visible wavelengths such as blue or green to avoid damaging the samples.)
5. Poly-L-lysine-coated glass slides were prepared as described elsewhere by incubating acid-washed slides in poly-L-lysine 0.1% solution for 30 min and drying them at 115°C for 30 min. Alternatively, poly-L-lysine-coated slides and cover slides can be purchased from different vendors (Polysciences Inc., BD Biosciences).
6. Although the 5:4:1 ratio works well for most LPSs, lipid A solution attained homogenization with less methanol. In general, the predominant factor to observe is the complete mixing of the phases independently of the ratio of solvents.
7. If necessary, further purification can be achieved through size exclusion chromatography. We used a Sephacryl HiPrep 16/60 (S-200 HR) column (50 × 1 cm) (GE Healthcare) for this purpose. The LPS–QDots conjugate was eluted in a narrow color band and stored in the dark at 4°C. Under these conditions, the LPS–QDots are stable for at least 1 month without any visible signs of flocculation or deterioration of fluorescent intensity. If such flocculation occurs, the constructs can be reconstituted by sonication.
8. DLS measures the size of particles, emulsions, and molecules in suspension, by means of the fluctuation of their Brownian

motion when the particles or molecules are illuminated with a laser. The intensity of the scattered light fluctuates at a rate that is dependent upon the size of the particles. Analysis of these intensity fluctuations yields the velocity of the Brownian motion and hence, the particle size using the Stokes–Einstein relationship.

Figure 2 shows the size distribution during the labeling process of smooth-type LPS from *E. coli* O55:B5, measured at 173° scattering and at room temperature (23°C). The experiment was performed with the following settings: material profile "protein," refractive index = 1.450, absorption = 0.001; dispersant water: temperature = 25°C, viscosity = 0.8872 cP, and refractive index = 1.330. The Mark–Houwink parameters are as follows: A parameter = 0.428, *K* parameter (cm^2/s) = $7.65.10^{-5}$; equilibration time = 2 min, and cell: ZEN0117-disposable micro-cuvette (100 μL).

Acknowledgments

The work was funded by the Arizona Translational Research Initiative Fund to Sergei Svarovsky. Dr. David Lowry from the School of Life Sciences Bioimaging Facility at Arizona State University is acknowledged for TEM assistance, and Dr. Jose Cano Buendia from the Biodesign Institute at ASU for tissue culture assistance.

References

1. Caroff M, Karibian D (2003) Structure of bacterial lipopolysaccharides. Carbohydr Res 338:2431–2447
2. David SA (2001) Towards a rational development of anti-endotoxin agents: novel approaches to sequestration of bacterial endotoxins with small molecules. J Mol Recognit 14:370–387
3. Wood SJ, Miller KA, David SA (2004) Anti-endotoxin agents. 1. Development of a fluorescent probe displacement method optimized for the rapid identification of lipopolysaccharide-binding agents. Comb Chem High Throughput Screen 7:239–249
4. Triantafilou K, Triantafilou M, Fernandez N (2000) Lipopolysaccharide (LPS) labeled with Alexa 488 hydrazide as a novel probe for LPS binding studies. Cytometry 41: 316–320
5. Pallarola D, Battaglini F (2008) An efficient method for conjugation of a lipopolysaccharide from *Salmonella enterica* sv. Minnesota with probes bearing hydrazine or amino functional groups. Anal Biochem 381: 53–58
6. Santos NC, Silva AC, Castanho M, Martins-Silva J, Saldanha C (2003) Evaluation of lipopolysaccharide aggregation by light scattering spectroscopy. Biophys J 84:345A
7. Luk JM, Kumar A, Tsang R, Staunton D (1995) Biotinylated lipopolysaccharide binds to endotoxin receptor in endothelial and monocytic cells. Anal Biochem 232:217–224
8. Troelstra A, Antal Szalmas P, de Graaf Miltenburg LAM, Weersink AJL, Verhoef J, Van Kessel KPM, Van Strijp JAG (1997) Saturable CD14-dependent binding of fluorescein-labeled lipopolysaccharide to

human monocytes. Infect Immun 65: 2272–2277

9. Resch-Genger U, Grabolle M, Cavaliere-Jaricot S, Nitschke R, Nann T (2008) Quantum dots versus organic dyes as fluorescent labels. Nature Methods 5:763–775
10. Dubertret B, Skourides P, Norris DJ, Noireaux V, Brivanlou AH, Libchaber A (2002) In vivo imaging of quantum dots encapsulated in phospholipid micelles. Science 298: 1759–1762
11. Anderson RE, Chan WCW (2008) Systematic investigation of preparing biocompatible, single, and small ZnS-capped CdSe quantum dots with amphiphilic polymers. ACS Nano 2:1341–1352

Chapter 11

Fluorescence-Based Methods to Assay Inhibitors of Lipopolysaccharide Synthesis

Marcy Hernick

Abstract

Treatment of infections caused by Gram-negative bacteria is difficult due in large part to problems arising from innate and acquired drug resistance, resulting in a limited number of effective antibiotics. Consequently, antibiotics that can circumvent mechanisms of drug resistance are needed. Lipid A is a glucosamine phospholipid that acts as an anchor for lipopolysaccharides (LPS) that comprise the outer membranes of Gram-negative bacteria, a barrier for small molecule entry into the cell, and is also the portion of LPS that stimulates the immune system in septic shock. Consequently, inhibitors of lipid A biosynthesis have the potential to function as antibiotics and/or anti-endotoxins in the treatment of Gram-negative bacterial infections. Current efforts in the development of antibiotics targeted against lipid A have focused on the metal-dependent deacetylase LpxC. Herein we describe fluorescence-based assays that can be used for the evaluation of LpxC inhibitors with the potential to serve as antibiotics.

Key words: Lipopolysaccharide, LPS, LpxC, Lipid A, Inhibitor, Metal-dependent deacetylase

1. Introduction

The outer membranes of Gram-negative bacteria are comprised of negatively charged lipopolysaccharides (LPS) that act as barriers to prevent the entry of small molecules, thereby contributing to the viability and innate resistance of these organisms (1, 2). LPS, also known as endotoxin, can structurally be broken down into three regions: O-antigen, core, and lipid A. The O-antigen and core portions of LPS are variable oligosaccharides, while lipid A is a relatively conserved glucosamine phospholipid that acts as an anchor for LPS and is the portion of LPS that stimulates the immune system in Gram-negative septic shock (3–5). Since the loss of lipid A results in decreased viability, increased sensitivity to

Otto Holst (ed.), *Microbial Toxins: Methods and Protocols*, Methods in Molecular Biology, vol. 739, DOI 10.1007/978-1-61779-102-4_11, © Springer Science+Business Media, LLC 2011

antibiotics, and decreased immune response in the case of septic shock, inhibitors of lipid A biosynthesis have the potential to function as antibiotics and/or anti-endotoxins (3–7).

Current efforts in the development of antibiotics targeted against lipid A have focused on the metal-dependent enzyme UDP-3-*O*-(*R*-3-hydroxymyristate)-*N*-acetyl-glucosamine deacetylase (LpxC) (6, 8–15). LpxC catalyzes the committed, and second overall, step in the biosynthesis of lipid A – the hydrolysis of UDP-3-*O*-(*R*-3-hydroxymyristate)-*N*-acetyl-glucosamine to form UDP-3-*O*-(*R*-3-hydroxymyristate)-glucosamine and acetate (16, 17). Metalloenzymes are attractive pharmaceutical targets because of past successes with metalloenzyme inhibitors (18–21). Inhibitors of metalloenzymes, such as LpxC, typically contain a metal-binding group that chelates the catalytic metal ion (i.e., hydroxamate, phosphonate, and carboxylate) (18–26). In order to pursue LpxC as a drug target, it is necessary to have assays that can readily evaluate potential LpxC inhibitors. Recently, several methods for assaying LpxC inhibitors were reviewed that predominantly relied on the use of radioactive materials (27). While these methods allow for a thorough evaluation of inhibitors, the disadvantages of these approaches are that they require the physical separation of molecules, and a license to work with radioactive materials. Therefore, they are not amenable to high-throughput evaluation of potential inhibitors and cannot be used in situations where working with radioactive materials is not possible. Consequently, fluorescence-based methods for the evaluation of potential LpxC inhibitors are needed.

Herein we describe fluorescence-based methods for the in vitro evaluation of LpxC inhibitors using purified recombinant LpxC. In these approaches, inhibitor affinity for LpxC is measured by monitoring changes in the fluorescent properties of small molecules. These methods utilize the fluorescent fatty acid analog 5-butyl-4,4-difluoro-4-bora-3a,4a-diaza-s-indacene-3-nonanoic acid (BODIPY® 500/510 C_4, C_9) that binds to the hydrophobic tunnel of LpxC, a critical region of the protein for high-affinity ligand binding (28). A similar approach can also be used to evaluate inhibitors with fluorescent properties using "direct" binding assays to measure inhibitor affinity, wherein the concentration of the small molecule is held constant and the concentration of LpxC is varied. For the evaluation of compounds with no inherent fluorescent properties, "competition-based" binding assays are used to measure inhibitor affinity for LpxC. In this approach, inhibitors that effectively compete with the BODIPY® fatty acid for binding to LpxC will displace the fluorescent molecule from the enzyme, resulting in a decreased fluorescence signal that can be correlated with inhibitor affinity.

The BODIPY®-based assays rely on changes in fluorescence signal, either fluorescence polarization/anisotropy (FP) or fluorescence intensity, to monitor ligand binding to LpxC and can accommodate different instruments by using different assay formats (cuvette, multi-well plate, and ultrafiltration). The general basis for FP-based assays is that the small fluorescent probe tumbles rapidly in solution and therefore, following excitation by polarized light, no polarized light is emitted. However, upon binding to an enzyme, the tumbling is slowed resulting in an increased polarization of the emitted light. The advantage of monitoring changes in FP is that it does not require separation of bound/unbound molecules; therefore, it is ideal for screening potential LpxC inhibitors using a high-throughput approach. Additionally, since FP is a ratiometric method, it is less sensitive to many fluorescent artifacts (29). If an instrument that can monitor FP is unavailable, the ultrafiltration method can be used to evaluate inhibitor affinity for LpxC (27, 28). This approach relies on the physical separation of free/bound ligand, followed by fluorescence intensity measurements to determine K_D values. While this approach is not ideal for HTS, it does offer an alternative to radioactivity-based assays for the evaluation of inhibitors. Herein we describe approaches to measure inhibitor affinity for LpxC using direct and competition-based assays using three different formats: cuvette, multi-well plate, and ultrafiltration.

2. Materials

2.1. LpxC Fluorescence Polarization-Based Binding Assay: Cuvette Method

1. Buffers and reagents: 200 mM bis-tris propane (Sigma, St. Louis, MO), 10 mM triscarboxyethylphosphine (TCEP; Gold Biotechnology, St. Louis, MO), pH 7.5, and 10 μM BODIPY® fatty acid (Invitrogen, Carlsbad, CA). Store all buffers and reagents at –20°C (see Notes 1–3).
2. LpxC (varied or constant concentration). Store LpxC at –80°C (see Notes 4 and 5).
3. Inhibitors (varied or constant concentration). Store at –20°C (see Note 6).
4. Costar 4826 (0.1–10 μL) tips, Costar 4863 (1–200 μL) tips, and Costar 3620 1.7-mL microcentrifuge tubes (Corning Incorporated, Corning, NY). 5-, 10-, and 50-mL polypropylene conical tubes (USA Scientific, Ocala, FL) (see Note 5).
5. Quartz fluorescence semi-microcuvettes (stirred cell) w/ microstir bar. Alternatively, standard quartz fluorescence semi-microcuvettes can be used in conjunction with a stirring rod or semi-microcuvette add-a-mixer (NSG Precision Cells, Farmingdale, NY) for sample mixing.

2.2. LpxC Fluorescence Polarization-Based Binding Assay: Multi-well Plate Method

1. Buffers and reagents: 200 mM bis-tris propane (Sigma, St. Louis, MO), 10 mM TCEP (Gold Biotechnology, St. Louis, MO), pH 7.5, and 10 μM BODIPY® fatty acid (Invitrogen, Carlsbad, CA). Store all buffers and reagents at −20°C (see Notes 1–3).
2. LpxC (varied or constant concentration). Store LpxC at −80°C (see Notes 4 and 5).
3. Inhibitors (varied or constant concentration). Store at −20°C (see Note 6).
4. Costar 4826 (0.1–10 μL) tips, Costar 4863 (1–200 μL) tips, and Costar 3620 1.7-mL microcentrifuge tubes (Corning Incorporated, Corning, NY). 5-, 10-, and 50-mL polypropylene conical tubes (USA Scientific, Ocala, FL) (see Note 5).
5. Corning 3650 nonbinding surface 96-well plates (Corning Incorporated, Corning, NY; 384-well plates can also be used) (see Note 7).

2.3. LpxC Fluorescence Intensity-Based Binding Assay: Ultrafiltration Method

1. Buffers and reagents: 0.1 M NaOH; 200 mM bis-tris propane (Sigma, St. Louis, MO), 10 mM TCEP (Gold Biotechnology, St. Louis, MO), pH 7.5; 10 μM BODIPY® fatty acid (Invitrogen, Carlsbad, CA). Store all buffers and reagents at −20°C (see Notes 1–3).
2. LpxC (varied or constant concentration). Store LpxC at −80°C (see Notes 4 and 5).
3. Inhibitors (varied or constant concentration). Store at −20°C (see Note 6).
4. Costar 4826 (0.1–10 μL) tips, Costar 4863 (1–200 μL) tips, and Costar 3620 1.7 mL microcentrifuge tubes (Corning Incorporated, Corning, NY). 5-, 10-, and 50-mL polypropylene conical tubes (USA Scientific, Ocala, FL) (see Note 5).
5. Microcons (MWCO 30K, Millipore, Bedford, MA). Microcons must be washed prior to assay to remove glycerol from the membranes (see Note 8).
6. Quartz fluorescence semi-microcuvettes or Corning 3650 nonbinding surface 96-well plates (384-well plates can also be used).

3. Methods

3.1. LpxC Fluorescence Polarization-Based Binding Assay: Cuvette Method

1. Mix together 100 μL 10× buffer, 10 μL 10 μM BODIPY® fatty acid, 100 μL inhibitor, and 790 μL water in a 1-mL stirred cell fluorescence cuvette. Stir at 30°C × 1 min.
2. Measure FP following Ex 480 nm and Em 516 nm (see Notes 9–11).

3. Add 10 μL LpxC to the solution. Stir/incubate at 30°C for 3 min (see Note 12).
4. Measure FP following Ex 480 nm and Em 516 nm (see Notes 11 and 12).
5. Repeat steps 2 and 3 for a total of ten additions of LpxC (100 μL total) (see Note 13).
6. Plot FP vs. [LpxC] concentration (direct binding assay, Fig. 1) or FP vs. [inhibitor] (competition-based binding assay, Fig. 2). K_D values are obtained by fitting Eq. 1 to these data for direct binding assays, or Eq. 2 for competition-based assays, where ΔFP is the change in FP observed; $FP_{initial}$ and +FPendpoint are the FP observed for the unbound fluorophore in the direct and competition-based binding assays, respectively; K_D^{FL} refers to the K_D value of the fluorescent analog binding to LpxC; and $K_D^{Inhibitor}$ refers to the K_D value of the unlabeled inhibitor binding to LpxC.

3.2. LpxC Fluorescence Polarization-Based Binding Assay: Multi-well Plate Method

1. Mix together 10 μL 10× buffer, 1 μL 10 μM BODIPY® fatty acid, 10 μL inhibitor, x μL LpxC, and 79 - x μL water in each well of plate (see Note 13).
2. Incubate at 30°C for 15–30 min to allow for ligand complex formation (see Note 12).
3. Measure FP following Ex 480 nm and Em 516 nm (see Notes 9–11).

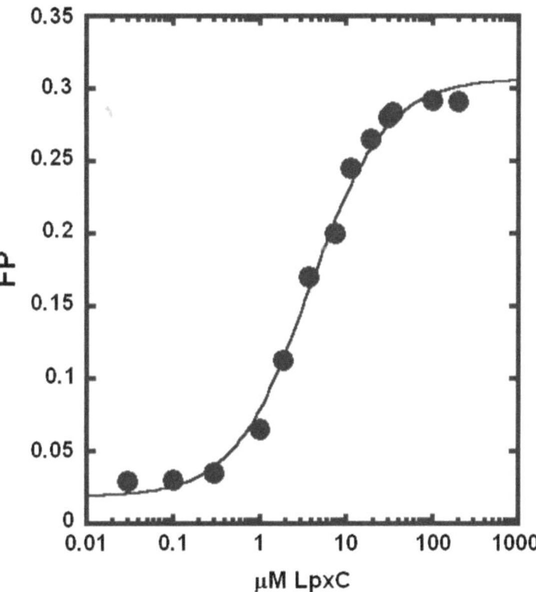

Fig. 1. Plot illustrating data analysis of a direct FP binding assay to monitor ligand binding to LpxC using hypothetical FP data.

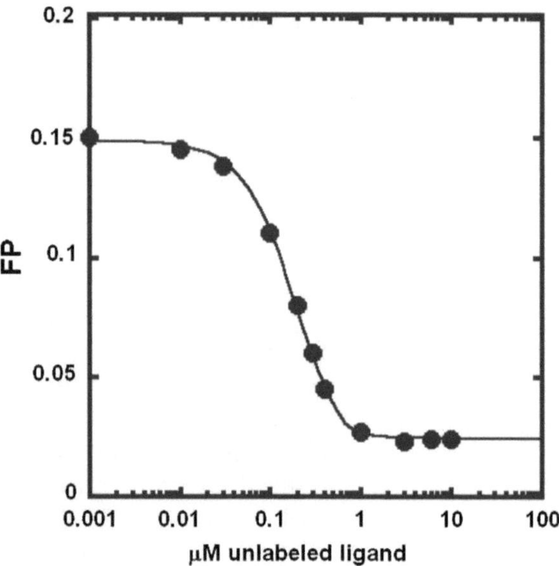

Fig. 2. Plot illustrating data analysis of a competition-based FP assay to monitor ligand binding to LpxC using hypothetical FP data.

4. Plot FP vs. [LpxC] concentration (direct binding assay, Fig. 1) or FP vs. [inhibitor] (competition-based binding assay, Fig. 2). K_D values are obtained by fitting Eq. 1 to these data for direct binding assays, or Eq. 2 for competition-based assays, where ΔFP is the change in FP observed; $FP_{initial}$ and $FP_{endpoint}$ are the FP observed for the unbound fluorophore in the direct and competition-based binding assays, respectively; K_D^{FL} refers to the K_D value of the fluorescent analog binding to LpxC; and $K_D^{Inhibitor}$ refers to the K_D value of the unlabeled inhibitor binding to LpxC.

3.3. Ultrafiltration Binding Assay

1. Mix together 50 µL 10× buffer, 30 µL 10 µM BODIPY® fatty acid, x µL LpxC, 50 µL inhibitor, and 370 - x µL H_2O in 1.7-mL microcentrifuge tubes. Spin in microcentrifuge (see Note 14).
2. Incubate at 30°C for 15–30 min to allow for ligand equilibration (see Notes 9 and 12).
3. Transfer the mixtures into pre-rinsed Microcons.
4. Spin at 3,000×*g* for 3 min. Transfer the filtrate back into sample reservoirs and mix samples. This step is to ensure that the filtrate sample is not diluted by residual buffer from the washing steps.
5. Spin at 3,000×*g* for ≤2.5 min. Dilute 75 µL of the filtrate with 925 µL 6 M urea. Transfer the diluted sample into a microvolume cuvette (1 mL) or a multi-well plate (100 µL) (see Note 15).

6. Flip sample reservoirs containing the retentate over new collection tubes. Spin at 3000 × *g* for 3 min. Dilute 75 μL of the filtrate with 925 μL 6 M urea. Transfer the diluted sample into a microvolume cuvette (1 mL) or a multi-well plate (100 μL) (see Note 15).
7. Measure fluorescence intensity following Ex 480 nm and Em 516 nm (see Notes 11 and 12).
8. Plot fluorescence vs. [LpxC] concentration (direct binding assay, Fig. 1) or fluorescence vs. [inhibitor] (competition-based binding assay, Fig. 2). K_D values are obtained by fitting Eq. 1 to these data for direct binding assays, or Eq. 2 for competition-based assays, where ΔFP is the change in FP observed, $FP_{initial}$ and $FP_{endpoint}$ are the FP observed for the unbound fluorophore, K_D^{FL} refers to the K_D value of the fluorescent analog binding to LpxC, and $K_D^{Inhibitor}$ refers to the K_D value of the unlabeled inhibitor binding to LpxC. Fluorescence intensity is substituted for FP in both equations.

$$FP_{Obs} = \Delta FP * \left(\frac{[LpxC]_{total}}{K_D^{FL} + [LpxC]_{total}} \right) + FP_{initial} \quad (1)$$

$$FP_{Obs} = \Delta FP * \left(\frac{[LpxC]_{total}}{[LpxC]_{total} + K_D^{FL} + [LpxC]_{total} * \left(1 + \frac{[Inhibitor]}{K_D^{Inhibitor}} \right)} \right) + FP_{endpoint} \quad (2)$$

4. Notes

1. This method can be used as a direct binding assay to determine K_D^{FL} values for fluorescent inhibitors, or as a competition-based assay with 5-butyl-4,4-difluoro-4-bora-3a, 4a-diaza-*s*-indacene-3-nonanoic acid (BODIPY® 500/510 C_4, C_9) to determine $K_D^{Inhibitor}$ values for inhibitors lacking fluorescent properties. For direct binding assays, [FL] (e.g., BODIPY® fatty acid or fluorescent inhibitor) is held constant ([FL]/K_D = 0.01) and [LpxC] is varied ([E]/K_D = 0.1–10) to maximize the observed signal change. Competition-based assays indirectly measure unlabeled inhibitor affinity by determining the ability of the inhibitor to compete with the BODIPY® fatty acid ligand for binding to LpxC ([BODIPY®] held constant, [FL]/K_D = 0.01). For competition-based binding assays, either [LpxC] or [inhibitor] can be varied and the

other reagent held constant. The [LpxC] should be varied for inhibitors that are precious and/or expensive, have low water solubility, or have the potential to form micelles. Otherwise, [inhibitor] can be varied and [LpxC] held constant ($[E]/K_D = 1$ to maximize signal change).

2. Water used in all the described methods is Milli-Q purified water.
3. For the BODIPY® fatty acid analog, prepare a 10 mM stock solution in DMSO (for solubility). Dilute 10 mM stock solution 1,000-fold with 1× buffer for 10 μM working stock solution. For direct binding assays on alternative fluorescent inhibitors, the BODIPY® fatty acid analog is omitted from the assay.
4. The concentration of LpxC required will depend on the type of binding assay (i.e., direct vs. competition based), and on inhibitor potency. Since LpxC overexpresses at high levels and is unstable at low concentrations (<1 μ*M*, (17)), the concentration of LpxC is typically varied in these experiments (typical range 1–200 μM). Both direct and competition-based assays require measurements at a minimum of six to ten different concentrations of LpxC (or inhibitor) to obtain a K_D value for a specific ligand.
5. LpxC is catalytically active with one bound Zn^{2+} (E·Zn complex). However, LpxC activity is inhibited upon binding of a second Zn^{2+} (E·Zn_2 complex) (17, 30). Therefore, low metal content tubes and tips are used throughout the methods to minimize Zn^{2+} contamination, and prevent the formation of E·Zn_2 complex during assays and storage. To minimize the formation of E·Zn_2, all metal ions should be removed from purified LpxC by treatment with chelating agents (20 mM dipicolinate, pH 7.5) prior to reconstitution with the catalytic Zn^{2+} (17).
6. Inhibitor stocks should be prepared at 10× the desired final concentration. Water-soluble LpxC inhibitors should be dissolved in 1× buffer. DMSO (≤10% v/v final) can be included for inhibitors with low water solubility. For experiments where [inhibitor] is varied, typically 6–10 different concentrations of inhibitor are used over the range around the expected K_D value ($[\text{inhibitor}]/K_D = 0.1$–10).
7. Nonbinding surface plates are used to prevent artifacts that can occur due to sample adherence to the sides of the wells.
8. To wash Microcons, fill sample reservoirs with 500 μL each of 0.1 M NaOH. Spin at 14,000×*g* for 12 min. Discard the filtrate. Repeat for a total of three washes. Fill the sample reservoirs with 500 μL 20 mM bis-tris propane and 1 mM TCEP, pH 7.5. Spin at 14,000×*g* for 12 min. Discard the filtrate.

Fill the sample reservoirs with 500 μL 20 mM bis-tris propane and 1 mM TCEP, pH 7.5. Microcons can be stored in buffer at 4°C for several days prior to use. Immediately before use, spin at 14,000 × *g* for 12 min. Insert the sample reservoirs over collection tubes. Spin at 3,000 × *g* for 3 min. Discard the filtrate. Place the sample reservoirs over an empty collection tube. Microcons are now ready for loading samples.

9. The incubation/equilibration time should be sufficient enough to ensure the E·L complex has reached equilibrium. The 3-min incubation time used here is more than sufficient for BODIPY® fatty acid analog binding to LpxC. For other fluorescent inhibitors, control experiments should be carried out wherein fluorescence measurements are made over time to determine the time taken to reach equilibrium (i.e., no additional change in fluorescence is observed).
10. Fluorimeter settings (slit width, averaging time) used will vary for each instrument and should be optimized. Initial values that can be used are a slit width of 10 (FP and fluorescence intensity readings), averaging time = 2 s (FP only), and three replicates for each point (FP only). The G-factor for the BODIPY® was determined to be ~1.89 (28). Since this value is unique for each fluorescent molecule, it should be determined separately for each fluorescent inhibitor that is used for (direct) FP-based binding assays.
11. The Ex 480 nm and Em 516 nm wavelengths correspond to those used to monitor the BODIPY® fluorophore. If alternative fluorescent inhibitors are used, the wavelengths must be changed to match those used for the desired fluorophore. In competition-based assays, control experiments that measure the fluorescence of inhibitors only (in 1× assay buffer) are used to ensure that the inhibitors being examined do not fluoresce at the wavelengths being used.
12. To ensure that the equilibrium is not perturbed by large changes in volume, it is important that the volume does not significantly change over the course of the experiment. For titration experiments, sufficient data can be obtained with a 10% change in volume, while ultrafiltration methods may require larger changes in volume to obtain sufficient sample. Samples should be spun for the shortest amount of time that provides an adequate signal change. Do not spin ultrafiltration devices for longer than 2.5 min in step 5 to ensure ≤20% change in volume.
13. [LpxC] in wells is varied. The amount of water per well is adjusted so that the total volume of samples in multi-well plates is held constant. The total volume for each well will depend on the plates that are being used. The 100 μL volume used here is for a 96-well half-area plate.

14. Control experiments should be carried out to ensure that the inhibitors do not stick to the Microcon membrane. Inhibitor-only solutions should be passed through the Microcons according to steps 1–7 (omit 6 M urea) and the concentration of inhibitor in filtrate and retentate measured. The concentration of inhibitor in the filtrate and retentate should be equal if there is no sticking to the membrane.
15. Samples are diluted in 6 M urea to denature the protein. This is to prevent artifacts that could arise due to fluorescence quenching/enhancement upon fluorophore binding to enzyme.

References

1. Raetz CRH (2002) Biosynthesis, secretion and function of Gram-negative endotoxin – a potent lipid activator of innate immunity. FASEB J 16:A521
2. Wyckoff TJO, Raetz CRH (1998) Altering the acyl chain specificity of UDP-GlcNac O-acyltransferases from *E. coli* and *P. aeruginosa*. FASEB J 12:L53
3. Raetz CRH, Whitfield C (2002) Lipopolysaccharide endotoxins. Annu Rev Biochem 71:635–700
4. Wyckoff TJO, Raetz CRH, Jackman JE (1998) Antibacterial and anti-inflammatory agents that target endotoxin. Trends Microbiol 6:154–159
5. Darveau RP (1998) Lipid A diversity and the innate host response to bacterial infection. Curr Opin Microbiol 1:36–42
6. Onishi HR, Pelak BA, Gerckens LS et al (1996) Antibacterial agents that inhibit lipid A biosynthesis. Science 274:980–982
7. Barb AW, Zhu P (2008) Mechanism and inhibition of LpxC: an essential zinc-dependent deacetylase of bacterial lipid A synthesis. Curr Pharm Biotechnol 9:9–15
8. Clements JM, Coignard F, Johnson I et al (2002) Antibacterial activities and characterization of novel inhibitors of LpxC. Antimicrob Agents Chemother 46:1793–1799
9. Kline T, Andersen NH, Harwood EA et al (2002) Potent, novel in vitro inhibitors of the *Pseudomonas aeruginosa* deacetylase LpxC. J Med Chem 45:3112–3129
10. Li XC, Uchiyama T, Raetz CRH, Hindsgaul O (2003) Synthesis of a carbohydrate-derived hydroxamic acid inhibitor of the bacterial enzyme (LpxC) involved in lipid A biosynthesis. Org Lett 5:539–541
11. Pirrung MC, Tumey LN, McClerren AL, Raetz CRH (2003) High-throughput catch-and-release synthesis of oxazoline hydroxamates. Structure activity relationships in novel inhibitors of *Escherichia coli* LpxC: in vitro enzyme inhibition and antibacterial properties. J Am Chem Soc 125:1575–1586
12. Pirrung MC, Tumey LN, Raetz CR et al (2002) Inhibition of the antibacterial target UDP-(3-O-acyl)-*N*-acetylglucosamine deacetylase (LpxC): isoxazoline zinc amidase inhibitors bearing diverse metal binding groups. J Med Chem 45:4359–70
13. Jackman JE, Fierke CA, Tumey LN et al (2000) Antibacterial agents that target lipid A biosynthesis in Gram-negative bacteria – inhibition of diverse UDP-3-O-(R-3- hydroxymyristoyl)-*N*-acetyglucosamine deacetylases by substrate analogs containing zinc binding motifs. J Biol Chem 275:11002–11009
14. Barb AW, Leavy TM, Robins LI et al (2009) Uridine-based inhibitors as new leads for antibiotics targeting *Escherichia coli* LpxC. Biochemistry 48:3068–3077
15. Barb AW, McClerren AL, Snehelatha K, Reynolds CM, Zhou P, Raetz CRH (2007) Inhibition of lipid A biosynthesis as the primary mechanism of CHIR-090 antibiotic activity in *Escherichia coli*. Biochemistry 46:3793–3802
16. Young K, Silver LL, Bramhill D et al (1995) The envA permeability cell division gene of *Escherichia coli* encodes the second enzyme of lipid A biosynthesis – UDP-3-O-(R-3-hydroxymyristoyl)-*N*-acetylglucosamine deacetylase. J Biol Chem 270:30384–30391
17. Jackman JE, Raetz CRH, Fierke CA (1999) UDP-3-O-(R-3-hydroxymyristoyl)-*N*-acetylglucosamine deacetylase of *Escherichia coli* is a zinc metalloenzyme. Biochemistry 38:1902–1911
18. Leung D, Abbenante G, Fairlie DP (2000) Protease inhibitors. Current status and future prospects. J Med Chem 43:305–341

19. Supuran CT, Scozzafava A, Casini A (2003) Carbonic anhydrase inhibitors. Med Res Rev 23:146–189
20. Supuran CT, Casini A, Scozzafava A (2003) Protease inhibitors of the sulfonamide type. Anticancer, antiinflammatory, and antiviral agents. Med Res Rev 23:535–558
21. Scozzafava A, Supuran CT (2000) Carbonic anhydrase and matrix metalloproteinase inhibitors: Sulfonylated amino acid hydroxamates with MMP inhibitory properties act as efficient inhibitors of CA isozymes I, II, and IV, and *N*-hydroxysulfonamides inhibit both these zinc enzymes. J Med Chem 43:3677–3687
22. White RJ, Margolis PS, Trias J, Yuan ZY (2003) Targeting metalloenzymes: a strategy that works. Curr Opin Pharmacol 3:502–507
23. Hernick M, Fierke CA (2005) Zinc hydrolases: the mechanisms of zinc-dependent deacetylases. Arch Biochem Biophys 433:71–84
24. Agrawal A, Romero-Perez D, Jacobsen JA, Villarreal FJ, Cohen SM (2008) Zinc-binding groups modulate selective inhibition of MMPs. ChemMedChem 3:812–820
25. Jacobsen FE, Lewis JA, Cohen SM (2007) The design of inhibitors for medicinally relevant metalloproteins. ChemMedChem 2:152–171
26. Shin H, Gennadios HA, Whittington DA, Christianson DW (2007) Amphipathic benzoic acid derivatives: Synthesis and binding in the hydrophobic tunnel of the zinc deacetylase LpxC. Biorg Med Chem 15:2617–2623
27. Hernick M, Fierke CA (2008) A Method to assay inhibitors of Lipopolysaccharide synthesis. In: Champney WS (ed) New antibiotic targets. Humana Press, Clifton, NJ, pp 143–154
28. Hernick M, Fierke CA (2006) Molecular recognition by *Escherichia coli* UDP-3-O-(R-3-hydroxymyristoyl)-*N*-acetylglucosamine deacetylase is modulated by bound metal ions. Biochemistry 45:14573–14581
29. Lakowicz JR (1999) Principles of fluorescence spectroscopy, 2nd edn. Kluwer Academic, New York
30. Whittington DA, Rusche KM, Shin H, Fierke CA, Christianson DW (2003) Crystal structure of LpxC, a zinc-dependent deacetylase essential for endotoxin biosynthesis. Proc Natl Acad Sci 100:8146–8150

Chapter 12

Micromethods for Lipid A Isolation and Structural Characterization

Martine Caroff and Alexey Novikov

Abstract

Lipopolysaccharides (LPSs) are major components of the external membrane of Gram-negative bacteria, and act as an effective permeability barrier. They are essentially composed of a hydrophilic polysaccharide region linked to an hydrophobic one, termed lipid A. Depending on their individual variable fine structures, they may be potent immunomodulators. Because of the structural importance and role of lipid A in bacterial pathogenesis, herein we describe two rapid practical micromethods for structural analysis. The first method allows the direct isolation of lipid A from whole bacteria cell mass; the second describes conditions for the sequential release of fatty acids, enabling the determination of their substitution position in the lipid A structure to be determined by matrix-assisted laser desorption/ionization mass spectrometry. Examples are given with reference to two major pathogens: *Bordetella pertussis* and *Pseudomonas aeruginosa*.

Key words: Gas chromatography, 3-Deoxy-D-*manno*-oct-2-ulosonic acid, Lipooligosaccharide, Lipopolysaccharide, Matrix-assisted laser desorption/ionization mass spectrometry, Tumor necrosis factor, Sodium dodecyl sulfate

1. Introduction

Endotoxins are major components of the Gram-negative bacterial outer membrane and occur either as lipooligosaccharides (LOS) composed of a lipid A region covalently linked to a core oligosaccharide, or as a lipopolysaccharide (LPS) composed of lipid A linked through a core oligosaccharide to a polysaccharide (O-chain) of repetitive oligosaccharide subunits. Lipid A is anchored in the outer layer of the external membrane. In some bacteria, it is a powerful immunomodulator, responsible for most of the biological activities of the LPS. Small amounts of LPS from such bacteria can have beneficial effects, but larger amounts may cause endotoxic shock.

Otto Holst (ed.), *Microbial Toxins: Methods and Protocols*, Methods in Molecular Biology, vol. 739,
DOI 10.1007/978-1-61779-102-4_12, © Springer Science+Business Media, LLC 2011

Lipopolysaccharide and lipid A preparations are variably heterogeneous. This is seen in their ladder-like staining on SDS–PAGE gels or on thin-layer chromatograms (TLC) (1), and in LPS, lipid A, and PS mass spectra (2, 3). This heterogeneity is due to different levels of biosynthesis related to different numbers of O-chain repetitive units. LPS molecular masses can range from about 2 to 20 kDa.

Analysis and comparison of the separated structural regions of LPS are usually performed on samples from preparative-scale extractions (2). However, the most commonly used methods for extracting both smooth- and rough-type LPSs are time consuming and require an appreciable amount of bacteria.

We present here a method for the direct selective separation and extraction of lipids A from bacterial cells, and for their analysis by mass spectrometry (4). The method, which is fast and easy, can be applied to microgram to milligram quantities of lyophilized bacteria, and is extremely efficient for monitoring lipid A structural modifications as a function of culture conditions, for example. Use of the direct lipid A isolation method will be illustrated by mass spectrometry analysis of *Pseudomonas aeruginosa* lipid A structural modifications induced by transitions between planktonic and biofilm growth styles of these bacteria (3).

Alkaline treatments of lipid A selectively remove ester-linked fatty acids (5, 6). Following the kinetics of fatty acid release, specific to each type (primary or secondary linked), by mass spectrometry we obtained more detailed structural information with respect to lipid A acylation patterns. This second method is perfectly compatible with the first and can be applied to micro amounts of lipid A directly isolated from bacteria, as illustrated using lipid A and LPS of *Bordetella pertussis*, the agent of whooping cough (7, 8) (see Notes 1 and 2).

2. Materials and Techniques

2.1. Bacterial Strains and Cultures

1. Grow Gram-negative bacterial cells in required experimental conditions.
2. Before harvesting, kill bacteria in cold 2% phenol or by incubation for 40 min at 56°C and then examine for the absence of viable bacterial growth.

2.2. Chemicals

Solvents: Ultrapure water was obtained with a Millipore Milli-Q® system (resistivity > 18 MΩ cm). Chloroform and methanol were of Normapur grade. *Acids*: Isobutyric acid was of synthesis grade and citric acid of Normapur grade. *Bases*: 28% Ammonia solution was of Normapur grade and 41% methylamine solution of purum grade. *MALDI–MS matrix*: 2,5-Dihydroxybenzoic acid (DHB) of purum grade was from Fluka.

2.3. Mass Spectrometry

Matrix-assisted laser desorption/ionization mass spectrometry (MALDI/MS) instructions can be followed using an appropriate MALDI/MS system.

The presented MS experiments were done in the linear mode, with delayed extraction, using a Perseptive Voyager STR (PE Biosystem, France) time-of-flight mass spectrometer. The reflectron mode can be used to obtain better resolution, but it could also lead to underestimation or even loss of some fragile molecular species (e.g., lipids A with substituted phosphate groups) due to their fragmentation between the ion source and the reflector. Lipid A and LPS mass spectra are recorded in the negative-ion mode at 20 kV with the adjustment of the extraction delay time to optimized resolution and signal-to-noise ratio. Spectra obtained in the positive-ion mode give important additional fragmentation data, but are not presented here (9).

Analyte ions are desorbed from the DHB matrix by pulses from a 337-nm nitrogen laser. Addition of citric acid at 0.1 M concentration to the matrix solution was found to improve the resolution and signal-to-noise ratio of LPS and lipid A mass spectra greatly (10). Citric acid chelates cations and causes disaggregation of LPS and lipid A micelles by insertion between the molecules, thus improving solubility and co-crystallization of the analyte with the matrix, resulting in a better desorption/ionization process.

3. Methods

Isolation of lipid A is effected by mild hydrolysis of the acid-labile link between the lipid moiety of LPS and the terminal Kdo residue of the oligo- or polysaccharide. The most widely used acetic acid hydrolysis method to liberate free lipid A can also cleave other acid-labile lipid A bonds, such as those linking glycosidic phosphate, ethanolamine pyrophosphate, amino sugars (e.g., aminoarabinose (5) or glucosamine (8)) esterifying lipid A phosphate groups. A later developed mild hydrolytic method using sodium acetate at pH 4.4 (11) was modified by the addition of 1% SDS, in order to disrupt micelles formed by amphiphilic LPS molecules (12). The modified method is presented here and used for comparison with the new micromethod (also shown below) (4). The latter procedure was found to be as mild as the former SDS-promoted hydrolysis and did not modify or eliminate any lipid A native element substitutions.

With the biological activities of LPS being intimately related to their structures (13, 14), it was a priority to find methods giving quick information on their structures, their stability and reproducibility with various growth batches, and their variability with their production under changing different growth conditions. As lipid A is responsible for most of the biological properties of the

LPS molecule, it was necessary to find a new lipid A isolation method that was fast, mild, and applicable to microscale bacterial samples. These conditions were achieved by hydrolyzing lipid A directly from the bacterial surface and using MALDI mass spectrometry for its characterization, i.e., establishing its molecular heterogeneity, degree of acylation, and the presence of some fatty acids and other substituents (4). Our aim was to develop methods for the production of sufficient material for several purposes in addition to direct MS analysis, e.g., analyses by chromatography and application of selective mild alkaline treatments for the sequential liberation of fatty acids in order to establish lipid A acylation patterns. The latter is illustrated here with micro quantities of *B. pertussis* lipid A directly isolated from bacterial cells. The lipid A molecules isolated by micro-hydrolysis of bacteria can thus be further characterized in a few hours as schematized in Fig. 1 with *B. pertussis* lipid A structure.

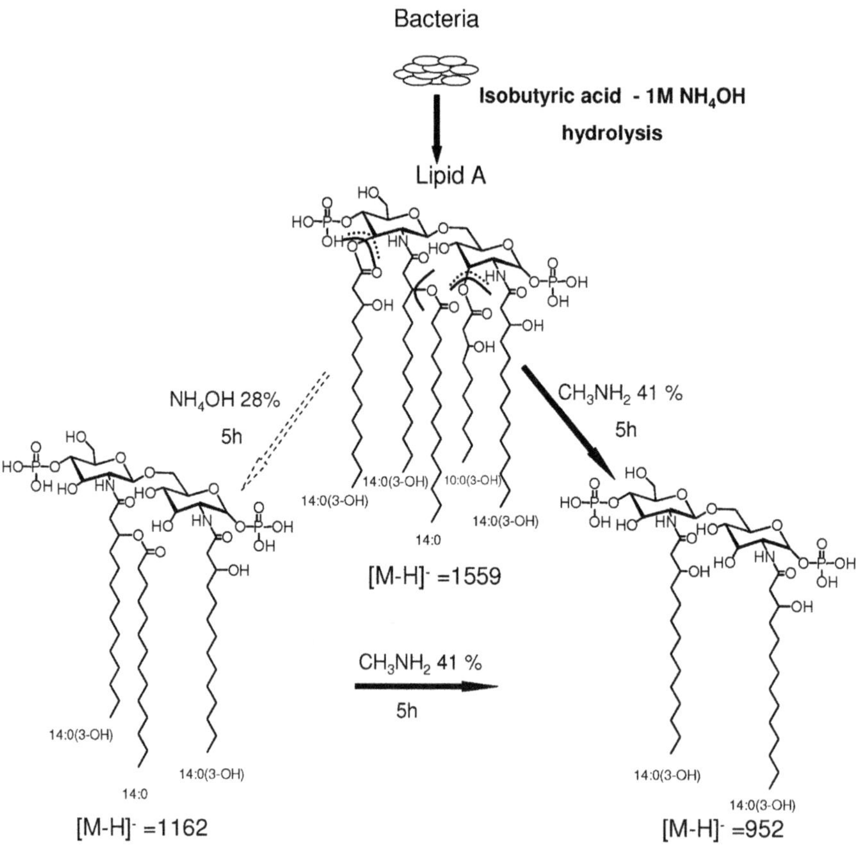

Fig. 1. Schematic representation of the analytical steps used for the analysis of lipid A isolated after hydrolysis of freeze-dried bacteria. The latter were hydrolyzed in a mixture of isobutyric acid–1 M ammonium hydroxide (5:3) for 2 h at 100°C. The *doted arrow* indicates alkaline treatment for 5 h at 50°C with 28% NH_4OH leading to partial *O*-deacylation, and the *black arrows* indicate alkaline treatment for 5 h at 37°C with 41% methylamine leading to complete *O*-deacylation. The structure displayed is that of *B. pertussis* lipid A.

3.1. Isolation of Lipid A from LPS on a Normal Analytical Scale (12)

1. Suspend 5 mg of LPS in 500 μL of 10 mM sodium acetate buffer containing 1% SDS (pH 4.4) in a screw-cap tube. Place in an ultrasound bath to obtain a clear solution.
2. Heat the sample at 100°C for 1 h. The duration can be adjusted depending on the solubility of samples (see Notes 3 and 4).
3. Cool in an ice-water bath and lyophilize the sample or dry it in a Speed Vac system.
4. Remove SDS by washing with a mixture of 100 μL of water and 500 μL of acidified ethanol (prepared by adding 100 μL of 4 M HCl to 20 mL of 95% ethanol). Centrifuge at 2,000 × *g* for 10 min (see Note 5).
5. Wash the sample twice again with 500 μL of non-acidified ethanol and centrifuge.
6. Lipid A is extracted from the dried sample by using 500 μL of a mixture of chloroform: methanol: water (3:1.5:0.25, by vol.).
7. Dry the lipid A extract, disperse it in 500 μL of water by sonication, and lyophilize to yield the lipid A as a powder.

3.2. Isolation of Lipid A from Whole Cells: A Micromethod (4)

1. After collecting a culture pellet, wash the cells, only if necessary, with a buffer or water and lyophilize it.
2. Under the hood, suspend 10 mg of lyophilized bacteria in 400 μL of a mixture of isobutyric acid:1 M ammonium hydroxide (5:3, v:v) in a screw-cap tube and heat for 2 h at 100°C with stirring (see Note 3).
3. Cool the sample to 4°C and centrifuge at 2,000 × *g* for 15 min.
4. Dilute the supernatant with water (1:1, v:v) and lyophilize it.
5. Wash the material obtained twice with 400 μL of methanol and centrifuge (2,000 × *g* for 15 min).
6. Extract the insoluble lipid A twice with 100 μL of chloroform:methanol:water (3:1.5:0.25, by vol.). This extract can be used directly for MALDI–MS.

3.3. Preparation of Isolated Lipid A for Mass Spectrometry (10)

1. Suspend lipid A at 1 μg/μL in a mixture of chloroform/methanol/water (3:1.5:0.25, by vol.) or use the lipid A extract obtained in Subheading 3.2 directly. Desalt the solution with a few grains of Dowex 50W-X8 (H^+) (see Note 6).
2. Deposit 0.5–1 μL of the solution on the target.
3. Add 0.5–1 μL of the matrix solution (DHB dissolved at 10 μg/μL in the same solvent, or in 0.1 M citric acid) and dry (see Notes 7 and 8).
4. Submit the sample on the target to MALDI–MS analysis, as described in Subheading 2.2 (see Note 9).

3.4. Sequential Liberation of Ester-Linked Fatty Acids of LPS or Lipid A by Mild Alkali Treatment for Analysis by Mass Spectrometry

Sequential fatty acid release by mild alkali treatment was found to be useful in establishing lipid A acylation patterns (5, 6). Ester-linked fatty acids in direct acylation of the di-glucosamine backbone (primary acylation) were released more readily than those bound to the hydroxyl groups of other fatty acids (secondary acylation). We discovered the experimental conditions for breaking: (1) only primary ester fatty acid bonds and (2) all fatty acid ester bonds. The conditions were established from our experience with numerous samples and proved that the technique can be scaled down if LPS or lipid A supply is limiting.

1. Lipid A solution (50 μL) obtained in Subheading 3.2, or the equivalent amount of LPS in an Eppendorf tube is dried under a stream of nitrogen or in a Speed Vac system.
2. In order to liberate only primary ester-linked fatty acids, add 50 μL of 28% ammonium hydroxide, sonicate, then close the tube, and stir for 5 h at 50°C in a Thermo-mixer system or in a simple thermostated bath using magnetic stirring.
3. Dry the sample under a stream of nitrogen.
4. Suspend the modified lipid A in 50 μL of a mixture of chloroform: methanol: water (3:1.5:0.25, by vol.) or in 50 μL of water when a LPS sample is treated. This sample is ready to be analyzed by MS, as described.
5. In order to liberate both the primary and secondary ester-linked fatty acids, repeat the above steps 1–4 replacing ammonium hydroxide by 41% methylamine and keeping the mixture for 5 h at 37°C under stirring.
6. If necessary, the liberated fatty acids can be recovered by extraction and tested by GC/MS.

4. Examples of Analysis

4.1. Illustration of P. aeruginosa Lipid A Structural Modifications Induced by the Switch Between Planktonic and Biofilm Lifestyles (3)

The micromethod for lipid A isolation, directly from bacterial cells, is particularly useful for comparing the lipid A structures of different species and strains of a genus, as well as that of the same bacteria grown under different culture conditions (e.g., media, temperature, oxygen tension, salt stress, bivalent ions concentration, etc.) or colonizing different niches (e.g., environmental isolates versus clinical isolates).

Here, we give the example of lipid A isolated from *P. aeruginosa* strain PAO1 grown in planktonic culture, in biofilm, and then in planktonic culture inoculated by biofilm isolated bacteria (hereafter assigned as "B→P" culture). Mass spectra of the lipid A extracted by the micromethod from ~4 mg samples of dried bacteria cultured in these three ways are presented in Fig. 2a–c. In all

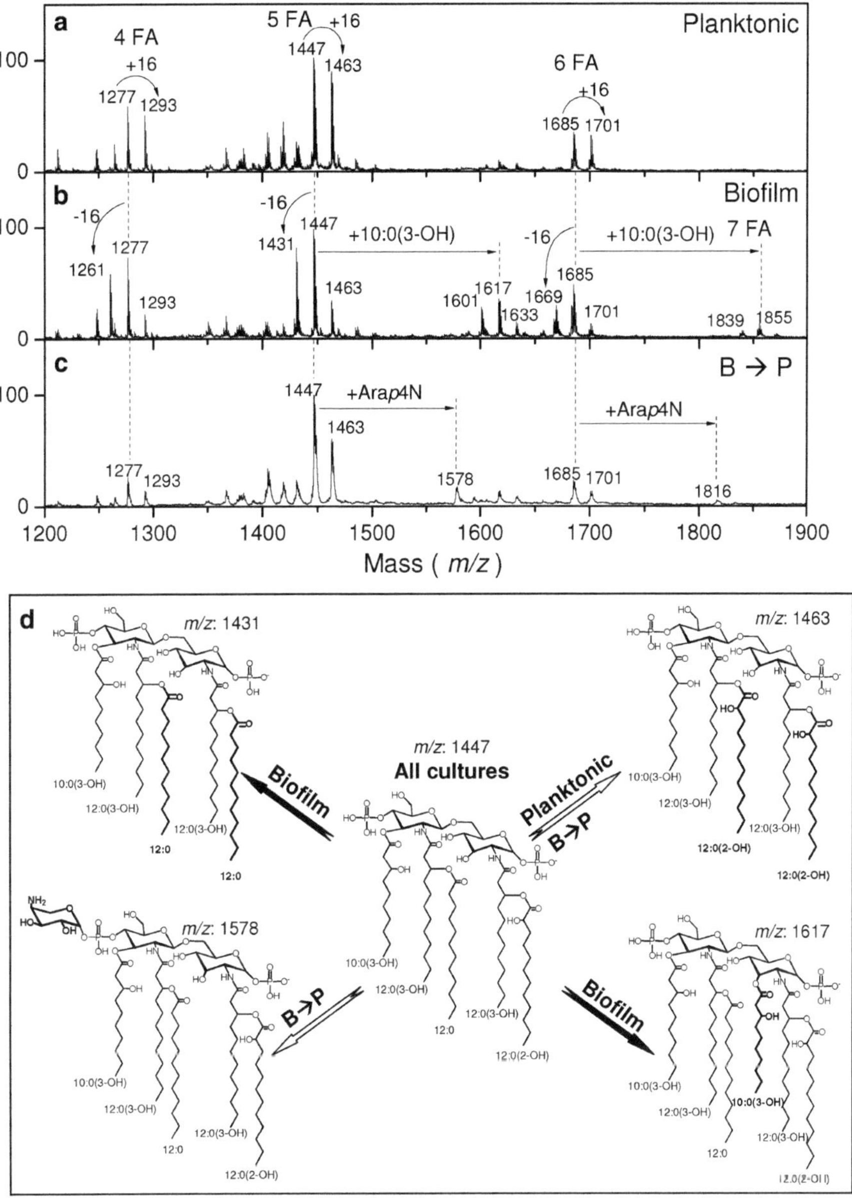

Fig. 2. Negative-ion MALDI mass spectra of lipids A isolated (**a**) from planktonic culture of PA01 bacteria, (**b**) from biofilm culture, and (**c**) from "biofilm to planktonic" culture. 4 FA: four fatty acids, 5 FA: five fatty acids, 6 FA: six fatty acids, 7 FA: seven fatty acids. (**d**) Structures of the major hexaacyl lipid A species appearing at *m/z* 1617 and of the different pentaacyl lipid A species appearing at *m/z* 1431, 1447, 1463, and 1578. Comparison of MALDI mass spectra shows an important decrease in 12:0(2-OH) content and an increase in 12:0 content in lipid A of biofilm-cultured bacteria.

the three spectra, three common major $[M-H]^-$ ion peaks were observed at *m/z* 1277, 1447, and 1685 corresponding to well-known tetra-, penta-, and hexaacyl molecular species whose structures were characterized and reported earlier (15). The structure of the common pentaacyl lipid A molecular specie (*m/z* 1447) is presented in Fig. 2d. The structures of the common tetra- and hexaacyl lipid A are derived from this structure, respectively, by the absence of a 10:0(3-OH) fatty acid at the C3′ position and by the addition of a C3′ acylhydroxyacyl palmitate (16:0). The appearance of other major peaks was shown to be dependent on culture conditions. Thus for planktonic and B→P cultures, major peaks at *m/z* 1293, 1463, and 1701 were observed corresponding to tetra-, penta-, and hexaacyl molecular species with two acyloxyacyl 12:0(2-OH) at C2 and C2′ positions (the structure of the corresponding pentaacyl lipid A is presented in Fig. 2d). In contrast, in the biofilm culture type, hydroxylation of secondary laurates was strongly attenuated, resulting in the appearance of major peaks at *m/z* 1261, 1431, and 1669 corresponding, respectively, to tetra-, penta-, and hexaacyl molecular species with two non-hydroxylated acyloxyacyl laurates (12:0) at C2 and C2′ positions (see Fig. 2d). In addition, for biofilm culture, we observed peaks corresponding to a hexa- (*m/z* 1601 and 1617) and a heptaacyl (*m/z* 1839 and 1855) molecular species containing 10:0(3-OH) fatty acid in the C3 position (see Fig. 2d). The latter two were absent in planktonic and B→P cultures. Finally, it was only for the B→P culture that we observed penta- and hexaacyl molecular species at *m/z* 1578, 1701, and 1816, in which one of the phosphate groups is substituted by aminoarabinose. The appearance and disappearance of peaks, specific for each culture lipid A product, were highly reproducible. In the context of lipid A biosynthesis, these data were obtained in a relatively short experiment time and with limited bacterial mass (4 mg of dried bacteria per sample). Thus using the new method, we were able to show that biofilm formation induces LPS structural modifications in *P. aeruginosa*, at the level of lipid A. These modifications contribute to higher inflammatory response from human monocytes (2).

4.2. Fatty Acids Positioning on the Di-glucosamine Backbone, by Sequential Alkaline Treatment (6): Example of B. pertussis Lipid A

The kinetics of esterified fatty acid liberation from lipid A samples was shown to be dependant on the fatty acid substitution on the di-glucosamine oligosaccharide backbone, and could be used to determine the fatty acid linkage. We first showed that liberation of primary ester-linked fatty acid (at C3 and C3′) was much faster than those of the fatty acids in secondary ester linkages (at C2 and C2′ acyloxyacyls), thus allowing a distinction between them by the application of the two-step protocol described in Subheading 3.4 (6). Furthermore, we observed that following kinetics of primary fatty acid liberation, we could discriminate between fatty acids located at the C3 and C3′ positions. We illustrate this new approach with *Bordetella* lipid A, whose structure is well characterized. The added interest of *B. pertussis* lipid A structure is that three different ester-linked fatty acids are present, hence offering selective liberation at the corresponding positions [14:0 – at C2′ acyloxyacyl, 14:0(3-OH) – primary at C3′ position, and 10:0(3-OH) – primary at the C3 position].

The kinetics presented in Fig. 2a was obtained from *B. pertussis* BP338 LPS sample (8) selected for its basically single lipid A molecular species composition that could offer a clear demonstration of our method. In the initial mass spectrum ($T = 0$), only one lipid A major peak is observed at *m/z* 1559 corresponding to the well-known pentaacyl molecular species structure displayed in Fig. 1. We followed the evolution of the mass spectra as a function of time during the alkaline treatment (28% NH_4OH at 50°C). Mass spectra were recorded after treatment times of 8 min, 15 min, 30 min, 1 h, and 2 h. Drastic differences in the fatty acid releasing times were observed that related to their linkage position. Thus, at C3, the 10:0(3-OH) is completely liberated only after 8–15 min, transforming the initial pentaacyl lipid A into a tetraacyl one (*m/z* 1389), while at C3′, the 14:0(3-OH) is released only after 1–2 h, leading to the disappearance of the peak at *m/z* 1389 and consecutive appearance of the one at *m/z* 1163. The time of release of the acyloxyacyl 14:0 at C2′ is again much longer. In fact, complete cleavage of this bond requires the use of stronger hydrolysis conditions as described in Subheading 3.4. Only after this second treatment, the peak at *m/z* 953 remains in the spectrum, corresponding to a diacyl lipid A with amide-linked 14:0(3-OH) fatty acids at C2 and C2′ (data not shown). Similar kinetics was obtained from a lipid A sample (Fig. 3b). In this case, we used lipid A from *B. pertussis* BP22 strain that was obtained by direct hydrolysis of dried bacterial cells. Two major molecular species were present in the initial spectrum at *m/z* 1559 (pentaacyl lipid A) and at *m/z* 1333 (tetraacyl lipid A with a free position at C3′). Despite this difference, as well as some very important differences in physical–chemical behavior between LPS and lipid A samples, the observed order in fatty acids release was the same. The first fatty acid to be released is at C3 with 10:0(3-OH) – in 30 min–1 h, then at C3′ with a 14:0(3-OH) – in 2–4 h, and finally at C2′ with the acyloxyacyl 14:0 – in excess of 4 h under these conditions as stronger alkaline conditions are required for its complete release. In conclusion, twice as long treatments were necessary to liberate the same fatty acids from the lipid A sample compared to LPS, probably owing to solubility difference between the two samples.

5. Notes

1. The methods presented here are rapid, facile, and effective procedures for the analysis of micro-scale samples and for following the evolution and biosynthesis of lipid A structures under different growth conditions, as illustrated. The method can also be readily adapted to the verification of quality and reproducibility of bacterial production.

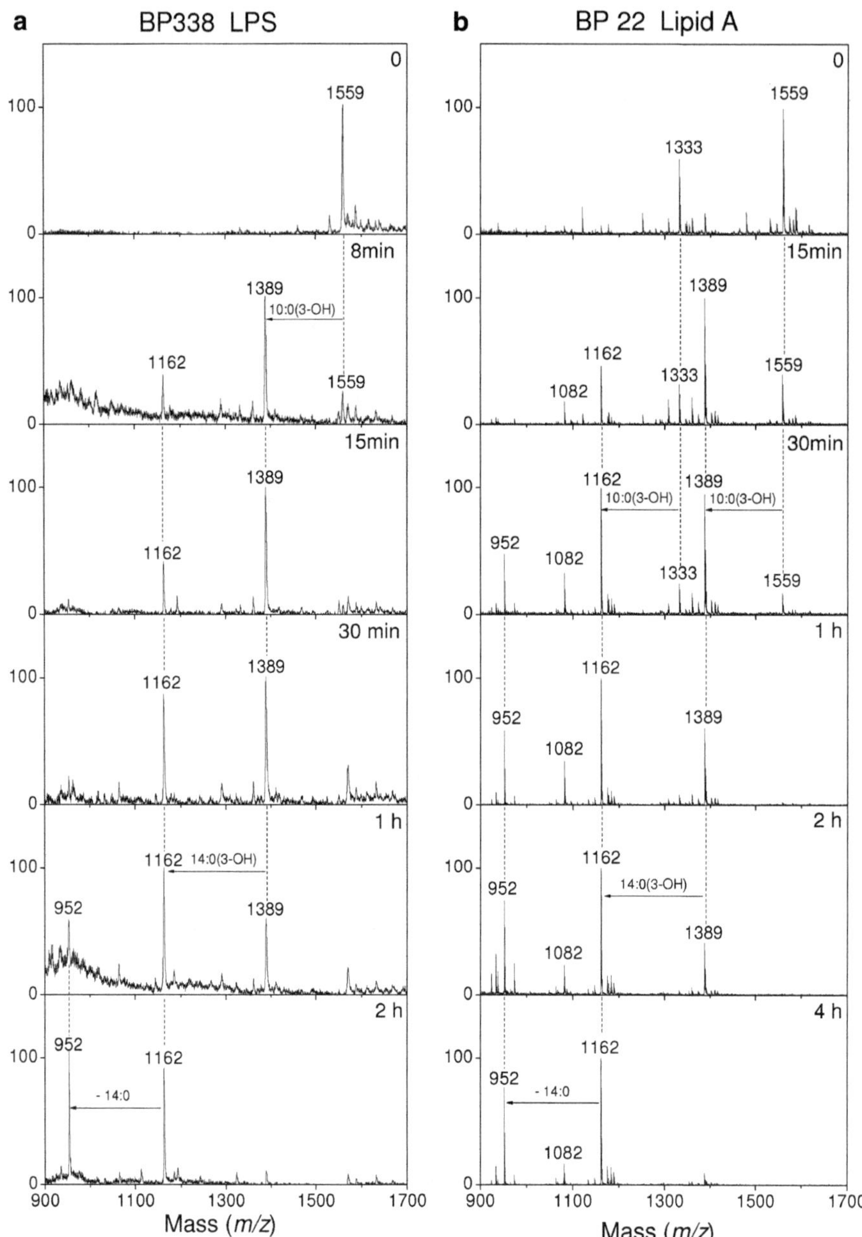

Fig. 3. Negative-ion MALDI mass spectra of lipid A obtained from (**a**) *B. pertussis* strain BP338 LPS and (**b**) *B. pertussis* strain BP22 lipid A. The corresponding time of hydrolysis with 28% ammonium hydroxide at 50°C is given in each spectrum. A clear difference is observed for fatty acid release between the two samples, due to differences in solubility. Both strain also display different degrees of heterogeneity naturally.

2. Each method can be experimentally adapted to the starting material and to the amount of bacterial sample.
3. Each hydrolysis time is given as the optimum for the large amount of different LPS samples tested; however, if some dephosphorylation is observed in the spectra, the acid hydrolysis time should be reduced and adjusted for a given LPS.

4. A phosphorylated terminal Kdo unit can render the glycosidic bond less labile (16). Depending on the mass ratio between the glycosidic and lipid moieties, the degree of acylation of the lipid A, and the presence of polar groups, the solubility of the samples can be highly variable. As with all methods, adjustments in conditions are sometimes required.
5. In the LPS hydrolysis method (Subheading 3.1), the ethanol supernatant should remain clear during the SDS extraction step, otherwise some lipid A might be lost in this fraction. If this is the case, ethanol should be concentrated and added back to the sample before *re-extraction*, and the centrifugation step should be performed at a higher speed and lower temperature.
6. The desalting process performed with the Dowex resin before MS analysis is a crucial step. It can be easily effected by the deposition of a drop of water on a piece of Parafilm on which a few grains of resin are added and mixed. Mixtures with chloroform are desalted in an Eppendorf tube.
7. During MS analysis, different ratios between sample and matrix volumes should be tested in order to obtain the best quality spectra with respect to signal-to-noise ratio and to the resolution of the peaks.
8. MALDI/MS is a very convenient MS process; however, depending on the temperature, humidity, or other variables, the crystallization can be a critical step and has to be repeated if a spectrum is not obtained.
9. Other MS methods can be successfully performed, but they are often more dependent regarding the presence of salt and other contaminants, thus requiring further purification steps.

Acknowledgments

Alexey Novikov is a recipient of a young researcher fellowship from INSERM (France). Part of this work was supported by the CNRS Group of Research GDR3048. This chapter is dedicated to Dr. Malcolm B. Perry (Ottawa, Canada) on his 80th birthday.

References

1. Caroff MG, Karibian D (1990) Several uses for isobutyric acid-ammonium hydroxide solvent in endotoxin analysis. Appl Environ Microbiol 56:1957–1959
2. Caroff M, Brisson J, Martin A, Karibian D (2000) Structure of the *Bordetella pertussis* 1414 endotoxin. FEBS Lett 477:8–14
3. Ciornei CD, Novikov A, Beloin C, Fitting C, Caroff M, Ghigo JM, Cavaillon JM, Adib-Conquy M (2010) Biofilm-forming *Pseudomonas aeruginosa* bacteria undergo lipopolysaccharide structural modifications and induce enhanced inflammatory cytokine response in human monocytes. Innate Immun 16(5):288–301

4. El Hamidi A, Tirsoaga A, Novikov A, Hussein A, Caroff M (2005) Microextraction of bacterial lipid A: easy and rapid method for mass spectrometric characterization. J Lipid Res 46:1773–1778
5. Aussel L, Therisod H, Karibian D, Perry MB, Bruneteau M, Caroff M (2000) Novel variation of lipid A structures in strains of different Yersinia species. FEBS Lett 465:87–92
6. Tirsoaga A, El Hamidi A, Perry MB, Caroff M, Novikov A (2007) A rapid, small-scale procedure for the structural characterization of lipid A applied to *Citrobacter* and *Bordetella* strains: discovery of a new structural element. J Lipid Res 48:2419–2427
7. Caroff M, Deprun C, Richards JC, Karibian D (1994) Structural characterization of the lipid A of *Bordetella pertussis* 1414 endotoxin. J Bacteriol 176:5156–5159
8. Marr N, Tirsoaga A, Blanot D, Fernandez R, Caroff M (2008) Glucosamine found as a substituent of both phosphate groups in *Bordetella* lipid A backbones: role of a BvgAS-activated ArnT ortholog. J Bacteriol 190:4281–4290
9. Karibian D, Brunelle A, Aussel L, Caroff M (1999) 252Cf-plasma desorption mass spectrometry of unmodified lipid A: fragmentation patterns and localization of fatty acids. Rapid Commun Mass Spectrom 13: 2252–2259
10. Therisod H, Labas V, Caroff M (2001) Direct microextraction and analysis of rough-type lipopolysaccharides by combined thin-layer chromatography and MALDI mass spectrometry. Anal Chem 73:3804–3807
11. Rosner MR, Tang J, Barzilay I, Khorana HG (1979) Structure of the lipopolysaccharide from an *Escherichia coli* heptose-less mutant. I. Chemical degradations and identification of products. J Biol Chem 254:5906–5917
12. Caroff M, Tacken A, Szabo L (1988) Detergent-accelerated hydrolysis of bacterial endotoxins and determination of the anomeric configuration of the glycosyl phosphate present in the "isolated lipid A" fragment of the *Bordetella pertussis* endotoxin. Carbohydr Res 175:273–282
13. Caroff M, Karibian D (2003) Structure of bacterial lipopolysaccharides. Carbohydr Res 338:2431–2447
14. Caroff M, Karibian D, Cavaillon JM, Haeffner-Cavaillon N (2002) Structural and functional analyses of bacterial lipopolysaccharides. Microbes Infect 4:915–926
15. Bedoux G, Vallee-Rehel K, Kooistra O, Zähringer U, Haras D (2004) Lipid A components from *Pseudomonas aeruginosa* PAO1 (serotype O5) and mutant strains investigated by electrospray ionization ion-trap mass spectrometry. J Mass Spectrom 39:505–513
16. Caroff M, Lebbar S, Szabo L (1987) Do endotoxins devoid of 3-deoxy-D-manno-2-octulosonic acid exist? Biochem Biophys Res Commun 143:845–847

Chapter 13

Two Efficient Methods for the Conjugation of Smooth-Form Lipopolysaccharides with Probes Bearing Hydrazine or Amino Groups. I. LPS Activation with Cyanogen Bromide

Fernando Battaglini and Diego Pallarola

Abstract

This chapter presents a conjugation method for coupling probes bearing hydrazine or primary amino groups to a smooth(S)-form lipopolysaccharide (LPS). LPS is modified by the activation of the hydroxyl groups present in its O-antigen moiety with cyanogen bromide in aqueous acetone. The method yields conjugates with good labeling ratios, preserving the endotoxic activity of the lipid A moiety. Conjugation of smooth-form LPS from *Salmonella enterica* sv. Minnesota with dansyl hydrazine and horseradish peroxidase yields labeling ratios above 300 nmol dansyl per mg LPS, with nearly no loss of the original endotoxin activity. In the case of horseradish peroxidase, introducing a spacer, a ratio of 28 nmol HRP per mg LPS is obtained, preserving 65% of the original endotoxic activity.

Key words: Lipopolysaccharide, Conjugation, Amino, Hydrazine, Hydrazide, Cyanogen bromide, Carbohydrate activation, Fluorescent probes, Enzymes

1. Introduction

Lipopolysaccharides (LPS) are complex negatively charged lipoglycans. Usually, they comprise three distinct regions: a fatty-acylated region called lipid A, responsible for its endotoxic effects; a short oligosaccharide which is the core region; and a polysaccharide which in most cases is an O-antigen portion with a composition that highly varies among Gram-negative bacteria (1). Looking at the structure of the molecule, the only functional group

Otto Holst (ed.), *Microbial Toxins: Methods and Protocols*, Methods in Molecular Biology, vol. 739,
DOI 10.1007/978-1-61779-102-4_13, © Springer Science+Business Media, LLC 2011

susceptible of being modified without severely loosing biological activity is the hydroxyl group, which is present in significant amounts in the O-antigen moiety. Fluorescein isothiocyanate has been used for such a purpose, but with a limited yield (2, 3).

Primary amines, hydrazine, and hydrazide functional groups are present in a myriad of fluorescent probes, spacers, and proteins. Therefore, the development of methods involving conjugation of LPS with molecules bearing these functional groups provides multiple labeling options, allowing the fine tuning of the properties of the selected probes. The simplest way to achieve this is through oxidation of the O-antigen moiety of LPS with sodium periodate, followed by the coupling containing a hydrazide probe, e.g., biotin-LC-hydrazide. Using this method, Luk et al. (4) obtained conjugates with a good endotoxic activity; however, the degree of labeling was low. Our group has synthesized biotin-labeled LPS by oxidation of the carbohydrate moiety with a good labeling ratio, yet with modest remnant endotoxic activity (5).

In order to maintain LPS molecular integrity, a milder method has to be used for the modification of the hydroxyl groups. Agarose activation with cyanogen bromide (CNBr) is a common practice for the further introduction of antibodies in the construction of affinity chromatography stationary phases. The activation through CNBr produces a reactive cyanate ester. It is important to control the conditions under which these reactions are carried out, as cyanate esters can be hydrolyzed to inert carbamates. For example, the use of NaOH to enhance the nucleophilicity of hydroxyl groups may lead to the formation of inert carbamates. Temperature control is another important factor in the activation process, as the activated molecule is very sensitive to this parameter. To overcome both these problems, a convenient activation method is the use of triethylamine (TEA) as a "cyano-transfer" agent in an aqueous acetone solvent at temperatures below 0°C, avoiding the need for a strong basic medium (6, 7). TEA reacts with CNBr to yield *N*-cyanotriethylammonium bromide (Scheme 1), which is a highly efficient activating agent for polysaccharides.

We have shown the efficient conjugation of smooth-form LPS from *Salmonella enterica* sv. Minnesota with dansyl hydrazine and horseradish peroxidase through primary amino groups. Using this technique, a labeling ratio of 330 nmol dansyl per mg LPS may be achieved, preserving practically all the original endotoxin activity. In the case of peroxidase, with previous addition of a spacer molecule, a labeling of 28 nmol HRP per mg LPS is obtained, preserving 65% of the original endotoxic activity (8).

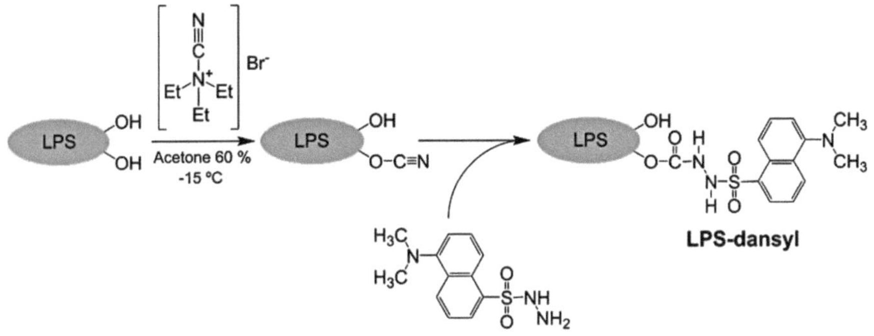

Scheme 1. Steps involved in the conjugation of dansyl hydrazine to LPS.

2. Materials

2.1. LPS Purification and Characterization

1. Smooth-form LPS from *Salmonella enterica* sv. Minnesota (LPS, Sigma, St. Louis, USA) (see Note 1). 50–60 mg LPS is dissolved in 3 mL of 10 mM Tris–HCl Buffer, pH 8.0, containing 0.2 M NaCl, 1 mM ethylenediaminetetraacetic acid, 0.02% (w/v) sodium azide, and 0.25% (w/v) sodium deoxycholate (*Buffer A*) (see Note 2).
2. *Buffer A* is used as elution buffer.
3. GE Amersham ÄKTA Explorer 10 FPLC system.
4. Sephacryl HiPrep 16/60 (S-200 HR, GE Healthcare, 120 mL capacity).
5. Desalting buffer: 15 mM phosphate buffer, pH 7.5, containing 15 mM NaCl.
6. 2 HiTrap desalting columns (GE Healthcare, 5 mL capacity each).
7. Standard RC dialysis tubing: 3500 MWCO Flat Width 18 mm (Spectra/Por, Spectrum Laboratories).
8. Lyophilizer.
9. Bio-Rad protein assay dye reagent concentrate.
10. Adenosine ribonucleic acid determined by UV spectroscopy ($A_{260\ nm}$ equal to 0.1 corresponds to 4 μg/mL of RNA (9)).
11. 3-Deoxy-D-*manno*-oct-2-ulosonic acid (Kdo, available from Sigma) (10).
12. Chromogenic LAL test (BioWhittaker).

2.2. LPS Activation with CNBr

1. 60% aqueous acetone.
2. 3 mg of purified LPS is dissolved in 4 mL of 60% aqueous acetone (see Note 3). Stir with vortex for 3 min and sonicate for 15 min to assure complete dissolution of LPS.
3. 1 M CNBr freshly prepared (see Note 4) in 60% aqueous acetone.
4. 0.15 M TEA in 60% aqueous acetone.
5. Refrigerated bath at –15°C (see Note 5).

2.3. Conjugation with Dansyl Hydrazine

1. 4 mg/mL of dansyl hydrazine (Fluka) in DMSO.
2. 1 M aqueous ethanolamine.
3. 2 HiTrap desalting columns (GE Healthcare, 5 mL capacity each).
4. Desalting buffer: 15 mM phosphate buffer, pH 7.5, containing 15 mM NaCl.
5. Standard RC dialysis tubing: 3,500 MWCO Flat Width 18 mm (Spectra/Por, Spectrum Laboratories).
6. Spectrophotometer Shimadzu UV-A110, dansyl $\varepsilon_{325\ nm} = 3{,}400\ M^{-1}\ cm^{-1}$.
7. PTI QuantaMaster UV–VIS spectrofluorometer, dansyl: $\lambda_{excitation} = 325$ nm; $\lambda_{emission} = 350–750$ nm.

2.4. Conjugation of Horseradish Peroxidase with FMOC-PEG Spacer

1. Dry acetonitrile (see Note 6).
2. 1.2 mg/mL FMOC-PEG-COOH (Scheme 2) (Novabiochem) in dry acetonitrile.
3. 5 mg/mL *N*-(3-dimethylaminopropyl)-*N*′-ethylcarbodiimide hydrochloride (EDC) in dry acetonitrile.
4. 2 mg/mL *N*-hydroxysuccinimide (NHS) in dry acetonitrile.
5. 10 mg/mL horseradish peroxidase (Biozyme, activity 260 U/mg measured by the pyrogallol assay (11)) in dry acetonitrile.
6. 0.3 M TEA in dry acetonitrile.
7. Desalting buffer: 15 mM phosphate buffer, pH 7.5, containing 15 mM NaCl.
8. 2 Hitrap desalting columns (GE Healthcare, 5 mL capacity each).
9. Standard RC dialysis tubing: 12,000 MWCO membrane, flat width 10 mm (Spectra/Por, Spectrum Laboratories).
10. Spectrophotometer Shimadzu UV-A110, FMOC $\varepsilon_{300\ nm} = 7{,}200\ M^{-1}\ cm^{-1}$.
11. 20% piperidine solution in DMF.
12. Silica on TLC aluminum foils with fluorescent indicator 254 nm.

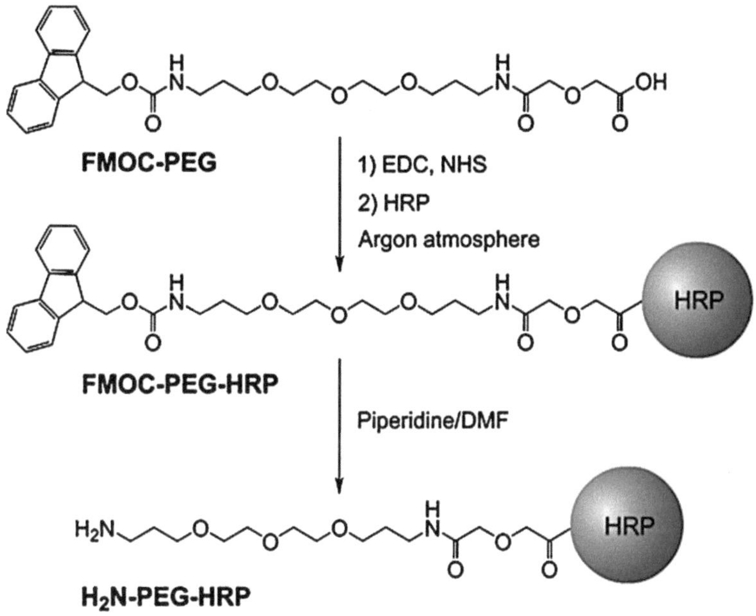

Scheme 2. HRP modification with PEG spacer.

2.5. Conjugation of PEG-HRP with Activated LPS (LPS-PEG-HRP)

1. 10 mg/mL PEG-HRP in 20% piperidine solution in DMF.
2. 1 M aqueous ethanolamine.
3. Amicon Millipore stirred cell 8010 with regenerated cellulose ultrafiltration membrane NMWL 30,000, 25 mm diameter.
4. 2 Hitrap desalting columns (GE Healthcare, 5 mL capacity each).
5. Desalting buffer: 15 mM phosphate buffer, pH 7.5, containing 15 mM NaCl.
6. ANX FF columns (high sub) (GE Healthcare, 1 and 20 mL capacity).
7. Binding buffer: 50 mM Tris–HCl buffer, pH 6.5.
8. Elution buffer: 50 mM Tris–HCl buffer, pH 6.5, containing 1 M NaCl.
9. Standard RC dialysis tubing: 12,000 MWCO membrane flat width, 10 mm (Spectra/Por, Spectrum Laboratories).
10. Spectrophotometer Shimadzu UV-A110, HRP $\varepsilon_{403\ nm} = 102{,}000\ M^{-1}\ cm^{-1}$.

2.6. Matrix-Assisted Laser Desorption–Ionization Mass Spectrometry

1. Sinapinic acid (Fluka) 1% (w/v) in an acetonitrile-0.1% trifluoroacetic acid (70:30, v/v) solution (photomatrix solution).
2. Protein samples (1 μM) prepared in Milli-Q water.

3. Bovine serum albumin (BSA) as external standard ($(M+H)^+ = 66{,}431$ and $(M+2H)^{2+} = 33{,}216$) (12).
4. Mass spectrometer (Omniflex Bruker Daltonics or similar).

2.7. SDS–Polyacrylamide Gel Electrophoresis

1. Separating buffer (4×): 1.5 M Tris–HCl buffer, pH 8.8, containing 0.4% SDS.
2. Stacking buffer (4×): 0.5 M Tris–HCl buffer, pH 6.8, containing 0.4% SDS.
3. 30% (w/v) acrylamide solution (2.6% C, 30.8% T) containing 0.8% (w/v) bisacrylamide (see Note 7).
4. *N*,*N*,*N*′,*N*′-Tetramethyl-ethylenediamine (TEMED).
5. Ammonium persulfate: 10% (w/v) solution in Milli-Q water. Store in single-use aliquots (200 μL) at −20°C.
6. Running buffer: 25 mM Tris–HCl buffer, pH 8.0, containing 0.19 M glycine and 0.1% (w/v) SDS.
7. Loading buffer (3×): 0.15 M Tris–HCl buffer, pH 6.8, 0.075% (w/v) bromophenol blue, 15% (w/v) 2-mercaptoethanol, 30% (v/v) glycerol, and 6% (w/v) SDS. Prepare loading buffer without 2-mercaptoethanol for samples detected by peroxidase activity assay.
8. Hoeffer SE-400 gel system.
9. Silver staining kit: *Solution A* contains 40% (v/v) ethanol and 5% (v/v) acetic acid. *Solution B* contains 40% (v/v) ethanol, 5% (v/v) acetic acid, and 0.7% (w/v) sodium periodate. *Staining solution*: 0.4% (v/v) NH_4OH, 20 mM NaOH, and 0.65% (w/v) $AgNO_3$ (store in the dark). *Developer*: 50 mg/mL citric acid and 0.5 mL/L 37% formaldehyde. All solutions are prepared in Milli-Q water.
10. Activity staining kit: 1.5% (v/v) aqueous Triton X-100. *Developer*: 100 mM phosphate buffer, pH 6, 0.6 mg/mL pyrogallol, and 22 mM H_2O_2.

3. Methods

3.1. LPS Purification and Characterization

1. Smooth-form LPS from *S. enterica* sv. Minnesota is chromatographically purified at 23–25°C with a column of Sephacryl HiPrep 16/60 in an ÄKTA Explorer 10 system. The column is equilibrated with at least 2 column volumes (CV) of elution buffer.
2. 3 mL of sample (50–60 mg) is injected and eluted isocratically at 0.5 mL/min during 1.5 CV and collected in 1-mL fractions. The purified LPS elute between fractions 35 and 68.

3. UV-detection at 260 and 280 nm is used to monitor the chromatographic separation (see Note 8).
4. Removal of the excess of detergent can be accomplished by injecting the fractions containing LPS into a series of two Hitrap desalting columns (5 mL capacity each); elution is carried out with 15 mM phosphate buffer, pH 7.5, containing 15 mM NaCl, at 2 mL/min elution rate (see Note 9).
5. The fractions are extensively dialyzed for at least 24 h against Milli-Q water using 3500 MWCO dialysis tubing at 4°C.
6. The dialyzed fractions are lyophilized and characterized by their content of proteins (Bio-Rad Protein Assay), adenosine ribonucleic acid, Kdo (10) (see Note 10), and endotoxic activity (LAL Test).
7. The fractions are stored as dried powder at –20°C until its use.

3.2. LPS Activation with CNBr

1. 4 mL of purified LPS solution is refrigerated under stirring in a refrigerated bath at –15°C (see Note 11).
2. Add 50 μL of 1 M CNBr and immediately 400 μL of 0.15 M TEA drop-wise during a period of 3 min.
3. The solution is left to react for 5 min under stirring and it is ready for further reactions (see Note 12).

3.3. Conjugation with Dansyl Hydrazine

1. To the activated LPS prepared in Subheading 3.2, 200 μL of 4 mg/mL of dansyl hydrazine is added at –15°C.
2. The solution is left to react for 2 h at 4°C under stirring in the dark.
3. Reaction is quenched by adding 1 mL of 1 M ethanolamine and left to react for at least 30 min.
4. Unreacted dansyl is removed by size-exclusion chromatography using a series of two 5 mL Hitrap desalting columns. The product is eluted using desalting buffer at a 2 mL/min rate measuring the absorbance at 325 nm.
5. The conjugate is dialyzed for 24 h against Milli-Q water using 3,500 MWCO dialysis tubing at 4°C in the dark.
6. The final product is lyophilized.
7. Dansyl content is established by fluorescence and UV–Vis spectrophotometry (see Note 13).
8. LPS content is established by Kdo analysis (see Note 10).
9. Endotoxic activity is determined by LAL test following the manufacturer instructions.

3.4. Conjugation of Horseradish Peroxidase with FMOC-PEG Spacer

1. To 5 mL FMOC-PEG-COOH solution, 1.2 mL of NHS and 2 mL EDC solutions are added under argon atmosphere. Leave the mixture to react under stirring for 1 h.
2. 5 mL of HRP and 0.25 mL of TEA solutions are added and left to react for 90 min under stirring.
3. Acetonitrile is removed by vacuum distillation at a temperature not higher than 25°C.
4. The dry product is dissolved in Milli-Q water and purified by size-exclusion chromatography using a series of two 5-mL HiTrap desalting columns. The reaction mixture is eluted using desalting buffer at an elution rate of 2 mL/min measuring the absorbance at 403 nm.
5. The purified product is dialyzed for 24 h against Milli-Q water using 3,500 MWCO dialysis tubing at 4°C in the dark and then lyophilized.
6. The modification of HRP can be checked by the appearance of a sharp band at 300 nm corresponding to the FMOC group (see Note 14).
7. FMOC protecting group is removed by dissolving the HRP derivative in a piperidine solution and left to react for 30 min at 25°C.
8. Removal of FMOC is followed by thin-layer chromatography (see Notes 14 and 15).

3.5. Conjugation of PEG-HRP with Activated LPS (LPS-PEG-HRP)

1. To the activated LPS prepared in Subheading 3.2, 1.5 mL of PEG-HRP obtained from the deprotection step is added at −15°C.
2. The solution is left to react for 2 h at 4°C under stirring.
3. The reaction is quenched by adding 1 mL of 1 M ethanolamine and left to react for at least 30 min.
4. The reaction mixture is concentrated to 3 mL employing a stirred cell with regenerated cellulose ultrafiltration membrane NMWL 30,000.
5. The concentrated product is injected into a series of two Hitrap desalting columns using binding buffer at an elution rate of 2 mL/min to eliminate low molecular mass compounds and to prepare the sample for the ionic chromatography purification step.
6. Unreacted peroxidase is separated from the conjugate by anionic chromatography using an ANX FF column (20 mL capacity) (see Note 16) at 23–25°C. 3 mL of unpurified LPS-PEG-HRP is loaded into the column. The chromatographic separation is performed at a flow rate of 4 mL/min. Binding buffer is applied to the column over 6 CV.

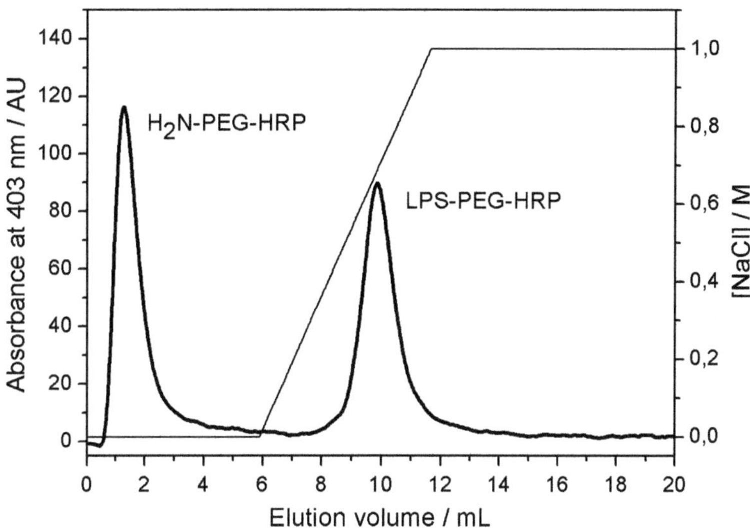

Fig. 1. Chromatogram (*bold line*) of the separation of LPS-PEG-HRP and unreacted H2N-PEG-HRP. The *thin line* represents the change in the NaCl concentration of the eluant. Column: ANX FF (high sub) (1 mL, GE Healthcare). Buffer elution: 1 mL/min (reproduced from ref. 7 with permission from Elsevier Science).

Elution is achieved with a gradient to elution buffer over 3 CV, from 0 to 100%, and maintained for 3 CV. The UV–Vis detector was set at 403 nm. An example of small scale purification is shown in Fig. 1.

7. The fractions corresponding to the conjugate are collected and concentrated as described in step 4.
8. The conjugate is desalted by repeating the step 5 using desalting buffer.
9. LPS-PEG-HRP solution is dialyzed for 24 h against Milli-Q water using 12,000 MWCO dialysis tubing at 4°C and lyophilized (see Note 17).
10. HRP content is established by measuring the absorbance at 403 nm while LPS content is established by Kdo analysis (see Note 10).
11. The conjugate is characterized by electrophoresis, enzymatic activity (11), and the LAL test for endotoxicity.

3.6. MALDI-TOF-MS

1. These instructions correspond to an Omniflex Bruker Daltonics mass spectrometer. The sample preparation protocol can be extended to any other spectrometer. The measurement conditions should be adapted according to the instrument employed.
2. Prepare the samples by mixing the protein solution with the photomatrix solution in a 1:10 (protein:matrix) ratio (protein mix).

3. Before applying the samples into the target plate, assure it is thoroughly cleaned. This is done by washing with methanol or isopropanol. If washing with the solvent is not sufficient, sonicate the target plate in methanol or isopropanol.
4. Apply 0.3 μL of the protein mix into at least three different spots and allow to air-dry. Repeat this step two times (see Note 18).
5. Follow the same procedure for the external standard, spotting around a protein sample in at least two different spots.
6. Set the mass spectrometer in linear mode and the linear detector under positive polarity scan mode with a 1.650 V gain.
7. Set the accelerating parameters to: Ion source 1 (IS1), 19 kV; IS2, 15 kV; and lens, 9.3 kV.
8. Ionization is achieved by irradiation with a nitrogen laser ($\lambda = 337$ nm) operating at 5 Hz with 200 ns of pulsed ion extraction delay.
9. Prior to acquisition of spectra, excess matrix is removed with 15 shots at a laser power of 80%. Each spectrum is recorded for the mass range of 10–100 kDa with 50 shots at a fixed laser power of 65% and maximum laser frequency.
10. Bruker flexAnalysis software is used to calibrate the spectrums by using the data collected from the BSA standard spectrum. Use the close neighbor BSA standard for calibration.

3.7. SDS–PAGE

1. These instructions correspond to a Hoeffer SE-400 gel system. They can be adapted to other formats.
2. Wash the glass plates with ethanol, rinse exhaustively with Milli-Q water, and let air-dry.
3. Prepare the 12.5% (see Note 19) polyacrylamide resolving gel solution by mixing 16.65 mL of acrylamide/bis solution, 10 mL separating buffer, 12.8 mL Milli-Q water, 20 μL TEMED, and 200 μL ammonium persulfate solution. Mix properly and pour the gel, leaving space for the stacking gel. Afterward, add ethanol over the resolving gel solution and let polymerize. The gel should polymerize for about 20 min at 25°C.
4. Pour off the ethanol and rinse the top of the gel with Milli-Q water.
5. Prepare the 5% polyacrylamide stacking gel solution by mixing 3.4 mL of acrylamide/bis solution, 2.4 mL stacking buffer, 14 mL Milli-Q water, 10 μL TEMED, and 100 μL ammonium persulfate solution. Mix properly, pour the stack, and insert the comb. The gel should polymerize within 15 min at 25°C.
6. After the stacking gel polymerized, remove the comb carefully and fill the wells with running buffer.

7. Add running buffer to the upper and lower chambers of the gel unit.
8. Prepare the sample by mixing one part of loading buffer with two parts of sample. Heat the mix for 5 min in a boiling water bath. Samples detected following peroxidase activity assay are just dissolved in loading buffer without 2-mercaptoethanol and are not boiled.
9. Load slowly 50 μL of each sample in a well. Sample mass should be between 1 and 5 μg for silver staining procedure, and between 5 and 15 μg for peroxidase activity assay.
10. Connect the gel unit to a power supply. Run the gel at 30 mA at 4°C until the tracking dye reaches the bottom of the gel. In these conditions, the run takes about 6 h. If cooling is not available, the gel should be run at lower current avoiding band spreading.
11. The gel is treated with one or both of the staining procedures (peroxidase activity assay is not compatible with previously silver stained gels).
12. Silver staining is carried out following the method described by Tsai and Frasch (13) (see Note 20).
13. Detection of peroxidase activity is carried out by rinsing the slab with 1.5% (v/v) aqueous Triton X-100 for 1 h at 4°C under smooth stirring. Then, wash twice with 1,000 mL Milli-Q water for 15 min each. Once the excess of detergent is removed, immerse the gel in the developer solution until the bands appear. This should occur in about 5 min at 25°C.

4. Notes

1. LPS is an endotoxin and pyrogen. It may cause fever and may be harmful by inhalation, ingestion, or skin absorption. A good laboratory technique should be employed; wear lab coat, gloves, and safety glasses. Work in a well-ventilated area and avoid contact with open wounds.
2. The complete dissolution of LPS may be accomplished by both stirring and sonicating. Avoid the excessive use of vortex in order to minimize the foam formation. Filtration with 0.45-μm pore size filter or centrifugation at 12,000 × g is recommended to separate the insoluble LPS before injecting into the column.
3. The fractions of purified LPS employed for the synthesis of conjugates may be chosen depending on the molecular mass and/or endotoxic activity that is needed for a specific purpose.

In our particular case, we selected the LPS that eluted between 41 and 47 mL (fractions 41–47, see Subheading 3.1). This fraction presents an endotoxic activity of 1.7 EU/ng, 0.3% RNA, and 0.3% proteins. Before purification, the LPS has 46% RNA and 0.8% proteins.

4. CNBr may be fatal if swallowed, inhaled, or absorbed through skin. Vapors cause severe irritation to eyes and respiratory tract, and cause burns to any area of contact. Contact with acids liberates poisonous gas. It also affects the blood, cardiovascular system, central nervous system, and thyroid gland. Impure material may explode. Wear lab coat and apron, gloves, and safety glasses. Work in fume hood, glove box, or ventilated cabinet. The contaminated material should be washed with 4% sodium hydroxide followed by 5% sodium hypochlorite followed by water. Dispose of waste solutions and materials appropriately.

5. Temperatures of the order of −15°C can be easily reached employing an ethanol–water bath. Any other solvent mix or a cryostat bath can be used to this purpose.

6. Acetonitrile may be dried by stirring with CaH_2 until no further hydrogen is evolved. The acetonitrile is then fractionally distillated under argon atmosphere and collected over Linde 4Å molecular sieves.

7. Acrylamide is a neurotoxin when unpolymerized. It is a possible human carcinogen and teratogen, and is readily absorbed through skin; therefore, handle it with care. Inhalation may be fatal. Wear lab coat, safety glasses, and gloves.

8. Since LPS does not present strong UV-absorbing groups, absorption of RNA at 260 nm, and proteins at 280 nm was followed during the elution. LPS is determined by its Kdo content and LAL test in each fraction.

9. When a series of two Hitrap desalting columns (5 mL each) is used, high molecular mass LPS elutes between 3 and 6 mL, if a 1-mL sample is injected. The sample volume can be scaled up to 3 mL maintaining good separation capacity, although eluting bands will appear wider. It is recommended to determine the chromatographic separation parameters by injecting a chemical analogous sample (i.e., labeled high molecular mass dextran with NaCl). Conductivity measurement provides a way to following the salt elution and allows estimating the retention volume of other colorless small molecules.

10. A known amount of the chromatographic LPS fraction used in the synthesis is strongly recommended as reference material to build the calibration curve. In this way, it is possible to establish a relationship between the concentrations of Kdo (10) and LPS.

11. The temperature for the activation of hydroxyl groups with the *N*-cyanotriethylammonium bromide complex may be kept low, avoiding the very rapid decomposition above –10°C (14).
12. In the conditions described, the activated LPS should be rapidly employed in the further reactions; otherwise, the global conjugation yield could be affected.
13. The determination of dansyl content in the LPS–dansyl conjugate may be accomplished by fluorescence spectroscopy. Build a calibration curve using standards of dansyl hydrazine from 0.5 to 12.0 μM in Milli-Q water. The LPS:dansyl ratio is calculated from samples of 5 μg/mL LPS–dansyl, below the critical micelle concentration (CMC) of LPS (15). At concentration below the CMC, both dansyl conjugate and free dansyl exhibited similar behavior, presenting an emission intensity that scales linearly with concentration (8).
14. The modification of HRP can also be followed by MALDI-TOF spectroscopy (8).
15. The deprotection reaction of FMOC-PEG-COOH can be monitored for the removal of FMOC group by employing silica TLC plates with ethyl acetate:dichloromethane (2:1, v/v) as solvent system. Detection of FMOC group is accomplished by using 254-nm UV light, whereas PEG can be detected by putting the dried TLC plate in a chamber with iodine crystals.
16. It is recommended to evaluate first the separating capacity of the column by performing analytical scale chromatographic runs. In this way, it is possible to determine which column is best suited to purify the sample. In the particular case presented in Subheading 3.5, we have used an ANX FF column (high sub) (1 mL). A volume of 100 μL of unpurified LPS-PEG-HRP (ca. 300 μg) was loaded into the column. Binding buffer was applied to the column over 6 CV. Elution was achieved with a gradient to elution buffer over 6 CV, from 0 to 100%, and maintained for 8 CV. The chromatographic separation was performed at 23–25°C at a flow rate of 1 mL/min, monitoring the absorbance at 403 nm.
17. The LPS-PEG-HRP solution could be stored at 4°C for 30 days without significant activity loss by supplementing the solution with 20% v/v glycerol and 10% (v/v) ethanol (be aware that the presence of glycerol lead to false positives results in the Kdo assay). Lyophilization is recommended when long storing periods are required.
18. An alternative sample application method consists in spotting the photomatrix solution and the protein solution in a sandwich-like procedure. First, 0.3 μL of the photomatrix solution is spotted to the target plate and let air-dry. Next, 0.3 μL

of the protein solution is applied and allowed to air-dry, followed by the spotting of 0.3 μL of photomatrix solution.

19. 12.5% polyacrylamide resolving gel is adequate for the LPS fraction employed in Subheading 3.5. For different size LPS, evaluate the use of gels with different crosslinking degree.

20. The washing steps are critical in order to obtain well-defined bands. After treating the gel with the oxidative solution, make sure to wash it extensively. Wash the gel with solution A for 10 min and with 40% ethanol for another 10 min under orbital stirring. Then, wash the gel three times with 1,000 mL Milli-Q water for 15 min each under orbital stirring. After treating with staining solution, repeat the Milli-Q water washing step. Remove the water and immerse the gel in developer solution. Observe the gel every 2 min, in order to avoid it get too dark. Be aware that the gel keeps getting darker even after it was removed from the developer solution. All steps should be performed in the absence of light.

References

1. Morrison DC, Ryan JL (eds) (1992) Bacterial endotoxic lipopolysaccharides. Vol I. Molecular biochemistry and cellular biology. CRC, Boca Raton
2. Troelstra A, Antal-Szalmas P, De Graaf-Miltenburg LAM, Weersink AJL, Verhoef J, VanKessel KPM, VanStrijp JAG (1997) Saturable CD14-dependent binding of fluorescein-labeled lipopolysaccharide to human monocytes. Infect Immun 65:2272–2277
3. Tobias PS, Soldau K, Gegner JA, Mintz D, Ulevitch RJ (1995) Lipopolysaccharide-binding protein-mediated complexation of lipopolysaccharide with soluble CD14. J Biol Chem 270:10482–10488
4. Luk JM, Kumar A, Tsang R, Staunton D (1995) Biotinylated lipopolysaccharide binds to endotoxin receptor in endothelial and monocytic cells. Anal Biochem 232:217–224
5. Priano G, Pallarola D, Battaglini F (2007) Endotoxin detection in a competitive electrochemical assay: synthesis of a suitable endotoxin conjugate. Anal Biochem 362:108–116
6. Kohn J, Wilchek M (1982) A new approach (cyano-transfer) for cyanogen-bromide activation of sepharose at neutral pH, which yields activated resins, free of interfering nitrogen derivatives. Biochem Biophys Res Commun 107:878–884
7. Wilchek M, Miron T, Kohn J (1984) Affinity chromatography. Methods Enzymol 104:3–55
8. Pallarola D, Battaglini F (2008) An efficient method for conjugation of a lipopolysaccharide from *Salmonella enterica* sv. Minnesota with probes bearing hydrazine or amino functional groups. Anal Biochem 381:53–58
9. Bergethon P (1998) The physical basis of biochemistry. Springer, New York, pp 254–256
10. Lee CH, Tsai CM (1999) Quantification of bacterial lipopolysaccharides by the purpald assay: measuring formaldehyde generated from 2-keto-3-deoxyoctonate and heptose at the inner core by periodate oxidation. Anal Biochem 267:161–168
11. Chance B, Maehly ACE (1955) Preparation and assays of enzymes. Methods Enzymol 2:773–775
12. Hirayama K, Akashi S, Furuya M, Fukuhara K (1990) Rapid confirmation and revision of the primary structure of bovine serum-albumin by ESIMS and FRIT-FAB LC MS. Biochem Biophys Res Commun 173:639–646
13. Tsai CM, Frasch CE (1982) A sensitive silver stain for detecting lipopolysaccharides in polyacrylamide gels. Anal Biochem 119:115–119
14. Fodor G, Abidi S, Carpenter TC (1974) N-cyanoammonium salts as intermediates in the von Braun cyanogen bromide reaction. J Org Chem 39:1507–1516
15. Aurell CA, Wistrom AO (1998) Critical aggregation concentrations of gram-negative bacterial lipopolysaccharides (LPS). Biochem Biophys Res Commun 253:119–123

Chapter 14

Two Efficient Methods for the Conjugation of Smooth-Form Lipopolysaccharides with Probes Bearing Hydrazine or Amino Groups. II. LPS Activation with a Cyanopyridinium Agent

Fernando Battaglini and Diego Pallarola

Abstract

This chapter presents a conjugation method for coupling probes bearing hydrazine or primary amino groups to a lipopolysaccharide (LPS). LPS is modified by the activation of the hydroxyl groups present in its O-antigen moiety with 1-cyano-4-dimethylaminopyridinium tetrafluoroborate (CDAP). The method yields conjugates with good labeling ratios, preserving the endotoxic activity of the lipid A moiety. Conjugation of smooth-form LPS from *Salmonella enterica* sv. Minnesota with dansyl hydrazine and horseradish peroxidase yields labeling ratios above 110 nmol dansyl/mg LPS, with nearly no loss of the original endotoxic activity. In the case of horseradish peroxidase, introducing a spacer, a ratio of 29 nmol HRP/mg LPS was obtained, preserving 65% of the original endotoxic activity and an enzymatic activity of 120 U/mg.

Key words: Lipopolysaccharide, Conjugation, Amino, Hydrazine, Hydrazide, 1-Cyano-4-dimethylaminopyridinium tetrafluoroborate, Carbohydrate activation, Fluorescent probes, Enzymes

1. Introduction

An efficient method for the conjugation of a lipopolysaccharide (LPS) with probes bearing primary amino and hydrazine moieties through activation with cyanogen bromide was introduced in Chapter 13. The reaction produces conjugates with an excellent ratio, the endotoxic activity of LPS is preserved, and it reveals the possibility of combining LPS with a myriad of probes bearing the aforementioned functional groups. In spite of all these advantages, the reaction conditions are quite unfavorable for further modification of LPS with an enzyme such as horseradish peroxidase (HRP).

Otto Holst (ed.), *Microbial Toxins: Methods and Protocols*, Methods in Molecular Biology, vol. 739,
DOI 10.1007/978-1-61779-102-4_14, © Springer Science+Business Media, LLC 2011

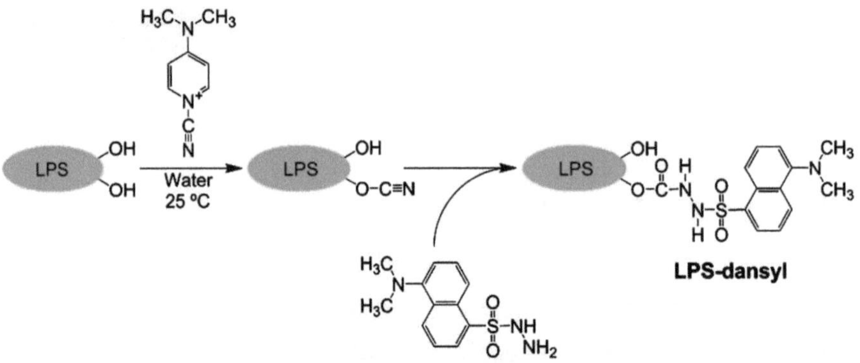

Scheme 1. Synthesis of CDAP-activated LPS and further modification with dansyl hydrazine.

Activation of hydroxyl groups in carbohydrates can also be achieved by treatment with 1-cyano-4-dimethylaminopyridinium tetrafluoroborate (CDAP) (Scheme 1) (1–5). In this case, the activation and further carbohydrate modification are produced under milder conditions in an aqueous solvent, that is, a friendlier environment for an enzyme. On the contrary, in an aqueous environment LPS forms micelles; to overcome this problem, different amphiphilic agents can be used to disaggregate LPS (6–9).

Our experience shows that the LPS conjugation in an aqueous solution can be successfully achieved in the presence of sodium deoxycholate (NaDC) (10). This chapter details with its conjugation using a fluorescent probe bearing a hydrazine group (dansyl hydrazine), and with an enzyme (HRP) using a primary amino group.

LPS from *Salmonella enterica* sv. Minnesota is conjugated with good labeling ratios with the fluorescent probe (110 nmol/mg), preserving 70% of its endotoxic activity. For the conjugation with HRP, diaminopolyethylene glycol (DAPEG) is used as a spacer. The spacer is bound to a periodate oxidized HRP (oHRP) and then reacted with the CDAP-activated LPS (Scheme 2). The conjugate LPS–DAPEG–oHRP presents a good conjugation ratio (29 nmol HRP/mg LPS) and preserves 65% of its original endotoxic activity, and the final enzymatic activity is 120 U/mg.

2. Materials

2.1. LPS Activation with CDAP

1. 2 mg/mL of purified LPS from *S. enterica* sv. Minnesota (Sigma, St. Louis, USA) prepared in Milli-Q water as described in Subheading 3.1 of Chapter 13.
2. 4.5 mM NaDC in Milli-Q water.
3. 2.5 mM EDTA in Milli-Q water.

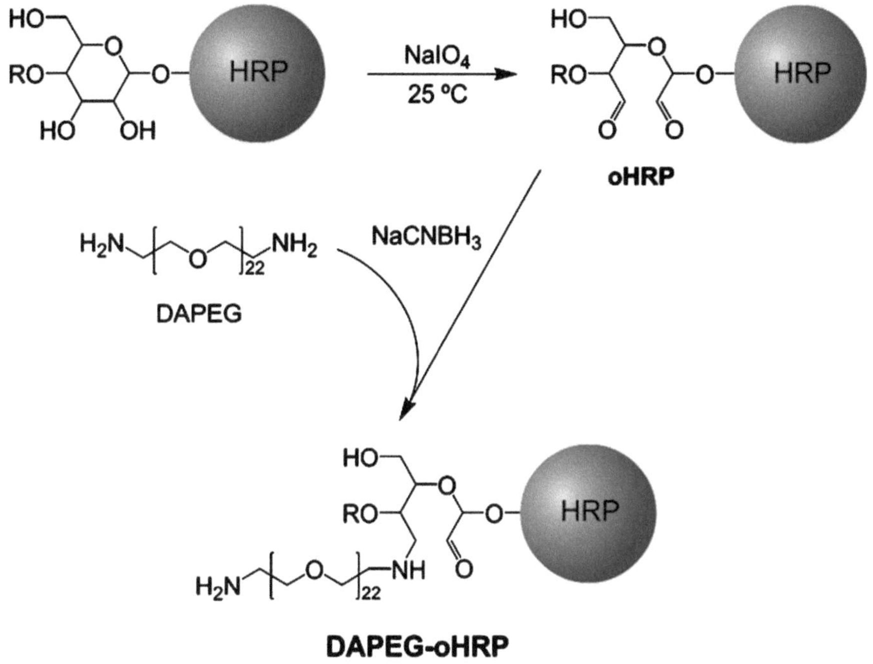

Scheme 2. Periodate oxidation of HRP and modification with diamino polyethylene glycol.

4. 0.2 M triethylamine (TEA) in Milli-Q water.
5. 100 mg/mL CDAP (Sigma-Aldrich) in acetonitrile.

2.2. Conjugation of LPS with Dansyl Hydrazine

1. 2.0 mg/mL dansyl hydrazine (Fluka) in 0.1 M sodium borate buffer, pH 9.3.
2. Ethanolamine.
3. Two HiTrap desalting columns (GE Healthcare, 5 mL capacity each).
4. Desalting buffer: 50 mM Hepes buffer, pH 7.5.
5. Standard RC dialysis tubing: 3,500 MWCO Flat Width 18 mm (Spectra/Por, Spectrum Laboratories).
6. Spectrophotometer Shimadzu UV-A110, dansyl $\varepsilon_{325\ nm} = 3{,}400\ M^{-1}/cm$.
7. PTI QuantaMaster UV–Vis spectrofluorometer, dansyl: $\lambda_{excitation} = 325$ nm; $\lambda_{emission} = 350–750$ nm.

2.3. Periodate Oxidation of Horseradish Peroxidase

1. 10 mg/mL HRP (Biozyme, activity 260 U/mg measured by the pyrogallol assay (11)) in Milli-Q water.
2. 50 mM $CaCl_2$ in Milli-Q water (see Note 1).
3. 100 mM sodium periodate in Milli-Q water.
4. Ethylene glycol.
5. Two HiTrap desalting columns (GE Healthcare, 5 mL capacity each).

6. Desalting buffer: 100 mM sodium bicarbonate, pH 9.5.
7. Bio-Rad protein assay dye reagent concentrate (Bio-Rad).

2.4. oHRP Conjugation with DAPEG

1. 34 mM DAPEG (MW 1,000 Da, provided by NOF, White Plains, NY, USA) solution in 100 mM sodium bicarbonate, pH 9.5.
2. Sodium borocyanohydride.
3. Ethanolamine.
4. Two HiTrap desalting columns (GE Healthcare, 5 mL capacity each).
5. Desalting buffer: 100 mM sodium bicarbonate, pH 9.5.

2.5. Conjugation of Activated LPS with DAPEG–oHRP

1. 12 mg/mL DAPEG–oHRP in 0.1 M sodium bicarbonate, pH 9.5.
2. Ethanolamine.
3. Amicon Millipore stirred cell 8010 with regenerated cellulose ultrafiltration membrane, NMWL 30,000, 25 mm diameter.
4. Two HiTrap desalting columns (GE Healthcare, 5 mL capacity each).
5. Desalting buffer: 50 mM Tris–HCl buffer, pH 6.5.
6. ANX FF anion-exchange column (high sub) (GE Healthcare, 20 mL capacity).
7. Binding buffer: 50 mM Tris–HCl buffer, pH 6.5.
8. Elution buffer: 50 mM Tris–HCl buffer, pH 6.5, containing 1 M NaCl.
9. Standard RC dialysis tubing: 12,000 MWCO membrane Flat Width 10 mm (Spectra/Por, Spectrum Laboratories).

3. Methods

3.1. LPS Activation with CDAP

1. 0.5 mL of 2 mg/mL purified LPS aqueous solution is vortexed for 3 min and sonicated for 15 min at 25°C (see Note 2).
2. 0.5 mL of 4.5 mM sodium deoxycholate is added.
3. 50 μL of 2.5 mM EDTA is added.
4. The solution is stirred for 30 min at 37°C, sonicated for 15 min, and again stirred for 30 min at 37°C.
5. 20 μL of 100 mg/mL CDAP in acetonitrile is added.
6. After 30 s, 20 μL of 0.2 M aqueous TEA is added.
7. The mixture is left to react for 150 s at 25°C under stirring. After this step, the LPS is modified with the chosen probe (see Note 3).

3.2. Conjugation with Dansyl Hydrazine

1. 500 μL of 2.0 mg/mL dansyl hydrazine in 0.1 M sodium borate buffer, pH 9.3, is added to the CDAP-activated LPS.
2. The mixture is left to react for 2 h in the dark at 25°C under stirring.
3. The reaction is quenched by the addition of 50 μL ethanolamine.
4. Unreacted dansyl is removed by size-exclusion chromatography using a series of two 5 mL HiTrap desalting columns. The product is eluted using desalting buffer at a 2 mL/min rate, measuring the absorbance at 325 nm.
5. The conjugate is dialyzed for 24 h against Milli-Q water using 3,500 MWCO dialysis tubing at 4°C in the dark.
6. The final product is lyophilized.
7. Dansyl and LPS content in the conjugate are determined following the procedure described in Subheading 3.3 of Chapter 13.
8. Endotoxic activity is determined by LAL test.

3.3. Periodate Oxidation of HRP

1. 10 mg/mL HRP is oxidized in 25 mM aqueous sodium periodate solution in the presence of 10 mM Ca^{2+} in the dark, under smooth stirring for 30 min at 25°C. Different concentrations of periodate were used in order to find the best conditions for the oxidation process (Fig. 1).
2. The reaction is quenched by the addition of 50 μL of ethylene glycol.
3. The periodate-oxidized enzyme (oHRP) is separated from the low molecular mass saccharides using two HiTrap desalting

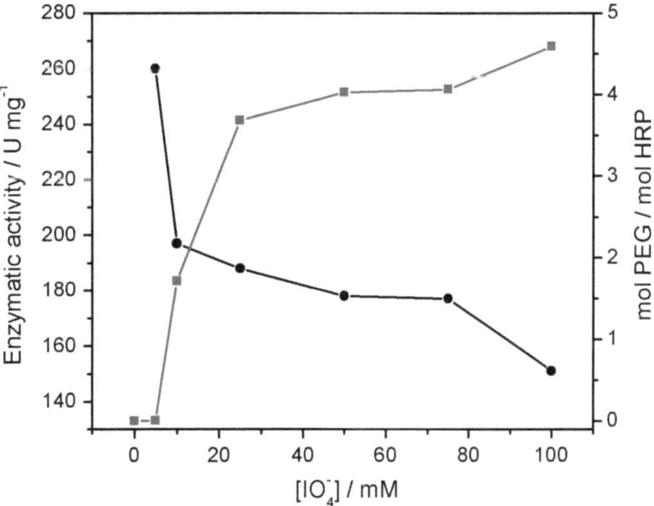

Fig. 1. Effect of the oxidative treatment on the enzymatic activity of HRP (*black line, circles*), and the DAPEG incorporation (*gray line, squares*) measured by MALDI-TOF-MS as described in Subheading 3.6 of Chapter 13.

columns in series using desalting buffer at an elution rate of 2 mL/min and measuring the absorbance at 403 nm. After this step, oHRP is in a suitable medium for coupling with the PEG spacer.

4. Enzymatic activity of the purified oHRP is determined using the pyrogallol method (11).
5. The molecular mass of oHRP can be determined by MALDI-TOF analysis as described in Subheading 3.6 of Chapter 13 (see Note 4).

3.4. oHRP Conjugation with DAPEG

1. 5 mg of sodium borocyanohydride is added to 0.5 mL of DAPEG 34 mM solution in 100 mM sodium bicarbonate, pH 9.5.
2. 1.5 mL of oHRP 3.5 mg/mL (prepared in Subheading 3.3) is added dropwise during a period not less than 3 min to the solution previously described.
3. The mixture is left to react in the dark for 2 h at 4°C under stirring.
4. The reaction is quenched with 50 μL of ethanolamine.
5. The enzyme is separated from the low molecular mass compounds as in Subheading 3.3 (step 3).
6. Peroxidase activity and MALDI-TOF-MS analysis are carried out as in Subheading 3.3 (steps 4 and 5, respectively).

3.5. Conjugation of Activated LPS with DAPEG–oHRP

1. To the CDAP-activated LPS prepared in Subheading 3.1, add 2 mL of DAPEG–oHRP obtained from the purification step described in Chapter 13.
2. The mixture is left to react for 2 h in the dark at 25°C under stirring.
3. The reaction is quenched with 100 μL of ethanolamine.
4. The product of the reaction is concentrated to 3 mL using a stirred cell with regenerated cellulose ultrafiltration membrane NMWL 30,000.
5. The concentrated product is injected into a series of two HiTrap desalting columns in series using desalting buffer at 2 mL/min elution rate, measuring the absorbance at 403 nm. In this way, the conjugate is in the right medium for the anionic purification step.
6. Unreacted peroxidase is separated from the conjugate by anion-exchange chromatography as reported in Chapter 13 (see Subheading 3.5).
7. Proceed with the steps described in Subheading 3.5 of Chapter 13.

4. Notes

1. Calcium chloride stabilizes the enzyme markedly since the integrity of active site depends on the presence of Ca^{2+} (12, 13).
2. The disaggregation of LPS is a critical step in its chemical modification. In order to obtain good conjugation yield, LPS must expose its polysaccharide domain to solution. The vortexing and sonicating procedures in addition to the use of amphiphilic compounds favor its monomerization.
3. In the conditions described, the activated LPS should be immediately employed in the further reactions; otherwise, the global conjugation yield could be affected.
4. The use of MALDI-TOF-MS for analyzing the mass change in HRP after the oxidative procedure is not fundamental. The periodate concentration for the oxidative process can be chosen from the data presented in Fig. 1.

References

1. Kohn J, Wilchek M (1983) 1-cyano-4-dimethylamino pyridinium tetrafluoroborate as a cyanylating agent for the covalent attachment of ligand to polysaccharide resins. FEBS Lett 154:209–210
2. Lees A, Nelson BL, Mond JJ (1996) Activation of soluble polysaccharides with 1-cyano-4-dimethylaminopyridinium tetrafluoroborate for use in protein–polysaccharide conjugate vaccines and immunological reagents. Vaccine 14:190–198
3. Bystricky S, Machova E, Bartek P, Kolarova N, Kogan G (2000) Conjugation of yeast mannans with protein employing cyanopyridinium agent (CDAP) – an effective route of antifungal vaccine preparation. Glycoconj J 17:677–680
4. Kossaczka Z, Szu SC (2000) Evaluation of synthetic schemes to prepare immunogenic conjugates of *Vibrio cholerae* O139 capsular polysaccharide with chicken serum albumin. Glycoconj J 17:425–433
5. Bystricky S, Paulovicova E, Machova E (2003) *Candida albicans* mannan-protein conjugate as vaccine candidate. Immunol Lett 85: 251–255
6. Olins AL, Warner RC (1967) Physicochemical studies on a lipopolysaccharide from cell wall of *Azotobacter vinelandii*. J Biol Chem 242:4994
7. McIntire FC, Sievert HW, Barlow GH, Finley RA, Lee AY (1967) Chemical physical and biological properties of a lipopolysaccharide from *Escherichia coli* K-235. Biochemistry 6:2363
8. Shands JW, Chun PW (1980) Dispersion of gram-negative lipopolysaccharide by deoxycholate – subunit molecular-weight. J Biol Chem 255:1221–1226
9. Panda AK, Chakraborty AK (1998) Interaction of mixed surfactants with bacterial lipopolysaccharide. J Colloid Interface Sci 203: 260–264
10. Pallarola D, Battaglini F (2009) Surfactant-assisted lipopolysaccharide conjugation employing a cyanopyridinium agent and its application to a competitive assay. Anal Chem 81:3824–3829
11. Chance B, Maehly ACE (1955) Preparation and assays of enzymes. Methods Enzymol 2:773–775
12. Ogawa S, Shiro Y, Morishima I (1979) Calcium-binding by horseradish peroxidase-C and the heme environmental structure. Biochem Biophys Res Commun 90:674–678
13. Laberge M, Huang Q, Schweitzer-Stenner R, Fidy J (2003) The endogenous calcium ions of horseradish peroxidase C are required to maintain the functional nonplanarity of the heme. Biophys J 84:2542–2552

Part III

Mold Fungus Toxins

Chapter 15

Extraction and Analysis of Fumonisins and Compounds Indicative of Fumonisin Exposure in Plant and Mammalian Tissues and Cultured Cells

Nicholas C. Zitomer and Ronald T. Riley

Abstract

Fumonisin mycotoxins are common contaminants in many grains, often at very low levels. Maize is particularly problematic as one of the organisms that commonly produce fumonisins, the fungus *Fusarium verticillioides*, often exists as an endophyte of maize. Fumonisin is a potent inhibitor of the enzyme ceramide synthase, and this inhibition results in the accumulation of a variety of upstream compounds, most notably, the sphingoid bases sphingosine, sphinganine, 1-deoxysphinganine and, in plants, phytosphingosine. Fumonisin exposure results in a wide variety of species, sex, and strain-specific responses. This method provides a relatively fast means of extracting fumonisins, sphingoid bases, and sphingoid base 1-phosphates from tissues and cells, as well as the subsequent analyses and quantification of these compounds using liquid chromatography/tandem mass spectrometry.

Key words: Fumonisins, Sphinganine, Sphingosine, Phytosphingosine, 1-Deoxysphinganine, Sphingoid base 1-phosphates, Liquid chromatography, Mass spectrometry

1. Introduction

The fungal genus *Fusarium* comprises organisms with an extremely diverse range of habitats and natural histories. *Fusarium verticillioides* is a notable member of the genus in that it commonly infects or inhabits maize tissues, and this infection/residence often coincides with its production of the fumonisin mycotoxins. Some of the diseases caused by *F. verticillioides* infection include seed rot, root rot, stalk rot, kernel, or ear rot, and seedling blight (1, 2). The production of fumonisins by *F. verticillioides* and other fumonisin-producing Fusaria results in contamination of maize destined for human and animal

Otto Holst (ed.), *Microbial Toxins: Methods and Protocols*, Methods in Molecular Biology, vol. 739,
DOI 10.1007/978-1-61779-102-4_15, © Springer Science+Business Media, LLC 2011

consumption. Fumonisins are known to cause a variety of diseases when ingested by animals (3–5), and have been implicated in human carcinogenesis (6) and neural tube defects (7), as well as plant disease (8, 9). There are many naturally produced forms of fumonisins, of which fumonisin B_1 is the most significant form based upon its toxicity and prevalence in the environment (10). While there are many other fumonisins (11), those most commonly found in maize are fumonisins B_1, B_2, and B_3 (10). Fumonisins of the B-series are known to disrupt sphingolipid metabolism in both plants and animals via inhibition of ceramide synthase, resulting in accumulation of free sphingoid bases and their 1-phosphates (3, 7, 8, 12–15). Recent advances have pointed to sphingoid bases and sphingoid base 1-phosphates as signaling molecules in plants and animals, as well as inducers of both increased proliferation and apoptosis (16–21). It was also shown recently that fumonisin production by pathogenic *F. verticillioides* strains is necessary for the development of leaf lesions in the sweet maize line "Silver Queen" (22). The ability to analyze plant and animal tissues for both fumonisin content as well as the results of fumonisin-induced disruption of sphingolipid biosynthesis is useful in order to investigate the toxicology of this important class of mycotoxins in animals, plants, and experimental models involving cultured cells.

The method described herein encompasses three major sections: the harvesting of tissues to be analyzed, extraction of relevant compounds from those tissues, and the analyses and quantification using liquid chromatography/tandem mass spectrometry (LC–MS). The section detailing harvesting procedures focus on the methodologies for collecting plant tissues, individual animal organs, and cells in culture. Each methodology has unique challenges, and the particular handling requirements for the different tissues vary greatly. The extraction procedures are fairly straightforward techniques, but vary based upon the tissue. The final section details how to utilize LC–MS for the analysis and quantification of fumonisin, sphingoid bases, and the sphingoid base 1-phosphates in a single analysis.

2. Materials

2.1. Chemicals/Reagents

1. Absolute ethanol (100%).
2. Acetonitrile (HPLC grade).
3. Ammonium formate (stock used at 15.75 g/L).
4. Chloroform (HPLC grade).
5. Drierite® (W.A. Hammond Drierite Co. Ltd., Xenia, OH, USA).

6. Formic acid (reagent grade >95%).
7. Methanol (HPLC grade).
8. Phosphate homogenization buffer – 0.05 M potassium phosphate, pH 7.0; 1.14 g/100 mL potassium phosphate dibasic (prepared fresh).
9. Phosphate buffered saline (PBS) (Dulbecco's PBS, Gibco/Invitrogen).
10. Water (HPLC grade).

2.2. Consumables

1. Disposable cell scrapers.
2. Glass test tubes (10–15 mL) with Teflon™-lined screw caps.
3. Imtact Scherzo 3 μm SM-C18 column, 150×3 mm (Imtact USA, Philadelphia, PA, USA).
4. Liquid nitrogen (industrial grade).
5. Paper bags (various sizes).
6. Pipette tips (various sizes).
7. Polypropylene microcentrifuge tubes (1.5 and 2.0 mL with snap tops).
8. Sample vials with Teflon™/silicone septa.
9. Solid phase extraction cartridges (C-18 Sep-Pak® Classic, Waters).
10. Spin-X™ centrifuge tube filters (0.45 μm nylon, COSTAR®, Corning Inc., Corning, NY, USA).

2.3. External and Internal Standards

1. Fumonisins B_1, B_2, and B_3 (PROMEC Unit, Tygerberg, Capetown, South Africa).
2. D-*erythro*-C_{17}-sphingosine-1-phosphate (C_{17}-So 1-P).
3. D-*ribo*-phytosphingosine-1-phosphate (Pso 1-P).
4. D-*erythro*-sphingosine-1-phosphate (So 1-P).
5. D-*erythro*-dihydrosphingosine-1-phosphate (Sa 1-P).
6. 1-Deoxysphinganine (1-deoxySa) (Avanti Polar Lipids, Alabaster, AL, USA).
7. Phytosphingosine (Pso, t18:0).
8. D,L-*erythro*-dihydrosphingosine (Sa, d18:0).
9. D-*erythro*-sphingosine (So, d18:1) (Sigma-Aldrich).
10. D-*erythro*-C_{16}-sphingosine (C_{16}-So, d16:1).
11. D,L-*erythro*-C_{20}-dihydrosphingosine (C_{20}-Sa, d20:0) (Matreya, Pleasant Gap, PA, USA).

2.4. Equipment

1. Automatic pipettes (various sizes).
2. Balances (0.01 mg–30 g).

3. Freeze dryer.
4. Freezers and refrigerators (4°C–80°C).
5. Platform rocker.
6. Sonicator with heating bath.
7. Vacuum desiccator.
8. Variable speed tissue homogenizer with Teflon™ pestle to fit 1.5-mL microcentrifuge tubes and with foot switch.
9. Vortex mixer.

2.5. Instrumentation

Finnigan Micro AS autosampler coupled to a Surveyor MS pump and Finnigan LTQ linear ion trap mass spectrometer (Thermo-Fisher, Woodstock, GA, USA) or instrumentation from different companies, but of equal value.

3. Methods

The method described herein has three major components: the harvesting of tissues to be analyzed, extraction of relevant compounds from those tissues, and the analyses and quantification by LC–MS. Briefly, the first component is the collection and storage of tissues or cells. Plant tissues are harvested and collected in paper bags and must be quickly transferred to a freezer (–20°C) until they can be freeze-dried (lyophilized) and stored in desiccators containing Drierite®. Animal tissues are typically collected at necropsy, and pieces of tissue of appropriate weight (50–100 mg fresh weight) for extraction should be quickly transferred to labeled polypropylene tubes cooled on ice and then subsequently transferred to a freezer (–20°C or lower). Attached cultured cells are typically collected by placing the culture flask or dish on ice after rinsing with cold PBS followed by removing the attached cells using a disposable cell scraper and collection of the detached cells in polypropylene tubes on ice. The detached cells are sedimented by centrifugation at 4°C and the supernatant is removed and the cell pellets are frozen (–20°C or lower). The second component of the method is the extraction of the compounds of interest. In the case of plant tissues, these are both the fumonisins and the sphingoid bases and sphingoid base 1-phosphates. Because fumonisins accumulate in plant roots, they are extracted quite easily along with the sphingoid bases and 1-phosphates (23). This is not the case in animal tissues or cultured cells. Typically in animal studies, the animals are fasted before necropsy and because fumonisins are poorly absorbed and rapidly eliminated, the levels in tissues are quite low and highly variable. Nonetheless, if the study does not require overnight fasting then fumonisins can be extracted and detected in tissues (24). In cultured cells, the detection of fumonisins is complicated by the fact that the extracellular concentration of fumonisins is typically extremely high compared to the intracellular concentration, and thus it is difficult to differentiate the residual extracellular fumonisin from the fumonisin that is truly inside the cell (see Note 1) (25).

The third component of the method is the analysis and quantitation which is similar for plant and animal tissues and cultured cells. Minor differences in the chromatography and mass spectrometry are necessitated by the fact that typically plant tissues are analyzed for both fumonisins and sphingoid bases and their 1-phosphates, whereas animal tissues and cultured cells are only analyzed for their sphingoid bases and sphingoid base 1-phosphates. Also, plant tissues from fumonisin-treated plants typically contain high levels of Pso, Sa, Pso 1-P, and Sa 1-P (23), whereas in mammalian tissues and cells treated with fumonisins Sa, Sa 1-P, 1-deoxySa, and to a much lesser extent So and So-1-P are the most commonly elevated (24, 26) (see Note 2).

3.1. Harvesting and Sample Preparation

The harvesting method employed concentrates on two major points. The first and most important aspect of this method is consistency. The scale of many experiments is often such that a harvesting step requires several people to help harvest in which many dozens of plants or animals need to be processed as quickly as possible. Due to the requirement of having many people involved in harvesting simultaneously, it is imperative that everyone knows the exact method prior to collection. This allows for the most consistent harvesting with the least variability introduced, and also allows for greater speed of harvest.

The main goals of sample preparation are to maintain consistency and keep everything cold throughout the period before tissues are freeze-dried or denaturing extraction solvents are added. This is especially important in the case of animal tissues and cultured cells, whereas the plant tissue can be handled at room temperature before they are frozen and after they have been freeze-dried. When samples are prepared fresh, it is very important to maintain the samples at low temperatures (4°C). At temperatures above 4°C, there is a greater chance of enzymatic conversion occurring among the compounds of interest. This is not a problem with the plant or animal tissues once they have been freeze-dried as long as they are kept desiccated. Once tissues are rehydrated, they must be kept cold (4°C).

3.1.1. Plant Tissues

1. Tissues should be collected as quickly as possible from the living plants. All aerial parts should be refrigerated (4°C) until all tissues have been collected, and then frozen at –20°C in paper bags to allow moisture to escape (not plastic bags).

2. Root tissues should be rinsed in 4–6 L of ice water. The most effective way to do this is by using a series of ice-water baths. First, remove as much soil as possible by gently massaging the root ball and then immerse the roots in the first ice-water bath and gently agitate until most of the soil is washed off (see Note 3). The roots are then transferred to the second ice-water bath where more detailed removal of soil from roots is performed. This is followed by one more ice-water bath aimed mainly at rinsing the most recalcitrant particles of soil away (see Note 4). The roots are then wrapped in an absorbent paper and frozen at –20°C.
3. These tissues are then freeze-dried to remove all water. Once freeze-dried, the tissues are transferred to glass test tubes with Teflon®-lined caps, then ground after cooling with liquid nitrogen until homogenized into a fine powder (see Note 5). This is achieved by submersing the bottom inch or so of the test tube containing the tissue in liquid nitrogen for a minute or two. After this, a metal spatula is inserted into the test tube and used to grind/pulverize the tissues into powder. Care must be taken to do this quickly so as not to allow for water to condense on the inside of the cold test tube.
4. The tubes are then placed, with tops loosened, in a plastic bag containing desiccant (Drierite® wrapped in a paper towel) overnight (at least 12 h) to remove any accumulated moisture.

3.1.2. Animal Organs

1. Pieces of tissues (50–100 mg) are weighed at necropsy, taking care to maintain the tissues on ice except when weighing. The weighed tissues are placed in labeled 1.5-mL conical polypropylene tubes and stored at –80°C.
2. Immediately before homogenizing, partially thaw tissues on ice and then add four volumes of freshly prepared cold phosphate homogenization buffer. The tissue is homogenized in 1.5 mL polypropylene tubes using a motor-driven foot switch-activated tissue homogenizer fitted with a Teflon™ pestle designed for use with the tubes. This process is conducted with the tube on ice.
3. A portion of the homogenate (100 μL = 20 mg wet weight) is then transferred to a cold 2-mL polypropylene tube. The remaining homogenates and the homogenate to be extracted (20 mg) can be stored at –80°C until ready for use.

3.1.3. Cultured Cells

1. After treatments, cells are washed twice with ice-cold PBS, and after removing most of the remaining PBS, the dishes or flasks placed on ice, and cells removed from the surface of the dishes using a disposable rubber scraper.

2. The detached cells are transferred with several PBS rinses to polypropylene tubes on ice and pelleted by low-speed centrifugation at 4°C. The PBS is removed and the cell pellets frozen at –20°C.

3.2. Sample Extraction

3.2.1. Plant Tissues

1. Approximately 10 mg of the finely ground freeze-dried tissue is added to a 2-mL polypropylene tube and the internal standards C_{16}-So and C_{17}-So 1-P (10 μL each at 100 ng/μL) are added. The samples are then stored in a vacuum dessicator containing Drierite® for 12–16 h.
2. An aliquot of 1:1 acetonitrile:water and 5% formic acid (1 mL/10 mg tissue) is added, the tubes briefly sonicated (this is just to wet and suspend the tissues which will then sink into the solvent) and then gently rocked on a platform rocker for 3 h (see Note 6).
3. The extracts are centrifuged at 15,000 × *g* and the supernatants filtered using 0.45-μm nylon centrifuge tube filters (COSTAR®, Corning Inc., Corning, NY, USA). The filtered extracts (100 μL aliquots) are diluted into the initial mobile phase (900 μL) used for LC–MS analysis (see below).

3.2.2. Animal Organs

1. The extraction mixture used is acetonitrile:water (1:1) with 5% formic acid (1.0 mL/20 mg tissue and containing C_{20}-Sa and C_{17}-So 1-P internal standards at a final concentration of 60 fmol/μL). Each polypropylene tube contains 20 mg of homogenized tissue in 100 μL phosphate homogenization buffer.
2. After adding the extraction mixture, the tubes are sonicated for 1 h at 50°C. Tubes are then mixed gently on a platform rocker for 3 h, followed by a quick vortexing.
3. The extraction mixture is centrifuged at 15,000 × *g* for 10 min, after which 400–500 μL of the supernatants are transferred to the 0.45-μm nylon centrifuge tube filters and clarified by centrifugation. The filtrates (250 μL) are then transferred to sample vials and 250 μL of acetonitrile:water (1:1, v/v) without formic acid but 10 mM ammonium formate is added.

3.2.3. Cultured Cells

1. Cell pellets are extracted with 1:1 acetonitrile:water (v/v) with 5% formic acid (400–1,000 μL/pellet) that contains C_{20}-Sa and C_{17}-So 1-P internal standards at a final concentration of 60 fmol/μl.
2. The tubes are sonicated at 50°C for 1 h and then placed on a platform rocker for 3 h, followed by centrifugation at 15,000 × *g* for 10 min. The supernatants (500 μL) are then transferred to 0.45-μm nylon centrifuge tube filters and clarified by centrifugation.

3.3. Analysis and Quantification

Stock solutions of fumonisins are dissolved in acetonitrile:water (1:1, v/v), and free sphingoid bases are dissolved in ethanol. Some sphingoid base 1-phosphates are not readily soluble in organic or aqueous solvents. In order to produce stock solutions, each sphingoid base 1-phosphate is first sonicated in a small volume (1% of the final volume) of formic acid and then dissolved in methanol, or in the case of Sa 1-P suspended in methanol (after the initial formic acid addition), followed by dissolution through the addition of chloroform (methanol:chloroform, 3:1, v/v) and by heating in warm water (see Note 7). The stock solutions are used to prepare working external standards (dissolved in initial mobile phase) for LC–MS and, in the case of C_{16}-So, C_{20}-Sa and C_{17}-So 1-P internal standards (dissolved in ethanol or the extraction mixture), for quantification.

Typical stock solutions and concentrations used to prepare working external standards are given below. For sphingoid base 1-phosphates, use of these stock solutions results in all of the sphingoid base 1-phosphates being at 0.3 μmol/mL, with the exception of the Sa 1-P, which is at 0.15 μmol/ml, as it is much less soluble than the other compounds (see Note 7). For the sphingoid bases, the preparations described below result in stock solutions of 1.5 μmol/mL.

3.3.1. Stock Preparation

Sphingoid Base 1-Phosphates

1. D-*erythro*-C_{17}-sphingosine-1-phosphate (C_{17}-So 1-P) prepared at 0.110 mg/mL in methanol with 0.1% formic acid.
2. D-*ribo*-phytosphingosine-1-phosphate (Pso 1-P) prepared at 0.120 mg/mL in 80% methanol, 20% chloroform with 0.1% formic acid.
3. D-*erythro*-sphingosine-1-phosphate (So 1-P) prepared at 0.114 mg/mL in 80% methanol, 20% chloroform with 0.1% formic acid.
4. D-*erythro*-dihydrosphingosine-1-phosphate (Sa 1-P) prepared at 0.057 mg/mL in 80% methanol, 20% chloroform with 0.1% formic acid.

Sphingoid Bases

1. Phytosphingosine (Pso) is prepared at 0.53 mg/mL in ethanol.
2. D,l-*erythro*-dihydrosphingosine (Sa) is prepared at 0.45 mg/mL in ethanol.
3. D-*erythro*-sphingosine (So) is prepared at 0.45 mg/mL in ethanol.
4. D-*erythro*-C_{16}-sphingosine (C_{16}-So) is prepared at 0.41 mg/mL in ethanol.

5. D,L-C_{20}-dihydrosphingosine (C_{20}-Sa) is prepared at 0.49 mg/mL in ethanol.
6. 1-Deoxysphinganine (1-deoxySa) is prepared at 0.43 mg/mL in ethanol.

3.3.2. Working Standards

1. Two separate standard mixtures are then made from these stock solutions; one standard of the sphingoid bases and one of the sphingoid base 1-phosphates. 20 μL of each sphingoid base stock is mixed and the resulting solution is brought to 1 ml with 1:1 water:acetonitrile with ammonium formate (20 μL of stock solution/ml of solvent; this results in a final concentration of 6.3 μg/mL, which is equivalent to 10 mM) and 5% formic acid.
2. The sphingoid base 1-phosphate stock is made by mixing 100 μL of each individual stock excepting Sa 1-P, which is added at 200 μL, and bringing the entire mixture up to 1 ml with with 1:1 water:acetonitrile with ammonium formate (10 mM) and 5% formic acid.
3. 10 μl of each of these two combined stocks are then added to 970 μL 1:1 water:acetonitrile with ammonium formate (10 mM) and 5% formic acid. To this is added 10 μL of a fumonisin standard mix containing FB_1, FB_2, and FB_3 all at 10 ng/μL. This results in a solution where all of the compounds present are at 300 fmol/μL except for the fumonisins, which are at 100 pg/μL.
4. This solution is then serially diluted, resulting in a standard curve consisting of solutions at 300, 30, and 3 fmol/μL, with fumonisins at 100, 10, and 1 pg/μL, respectively.
5. The sphingoid base and sphingoid base 1-phosphate curves are used only to identify retention time and fragmentation patterns of the compounds of interest. The fact that the curve is used is only to ensure that the response to these compounds remains linear over the concentrations included.
6. The sphingoid bases and sphingoid base 1-phosphates are quantified by comparing the response of each compound to that of the internal standard (C_{16} or C_{20} in the case of the sphingoid bases, and C_{17} in the case of the sphingoid base 1-phosphates).
7. The fumonisins are quantified against the external standard curve (i.e. the area under the curve is evaluated with regards to a best fit line of concentration versus area response, performed within the Xcaliber software used to operate the instrumentation).

3.3.3. LC–MS Sample Analysis

1. Analysis by LC–MS is conducted using a Finnigan Micro AS autosampler coupled to a Surveyor MS pump (Thermo-Fisher, Woodstock, GA, USA). Separation is accomplished

using an Imtact Scherzo 3 μm SM-C18 column, 150 × 3 mm (Imtact USA, Philadelphia, PA) (see Note 8).

2. The sample tray in the autosampler is cooled to 12°C (see Note 9). The flow rate is 0.2 ml/min, and the initial mobile phases used are 97% acetonitrile/2% water/1% formic acid (solvent A) and 2% acetonitrile/97% water/1% formic acid (solvent B).
3. For plant tissues, the gradient runs from 30% A, 70% B to 50% A, and 50% B at 7 min. The gradient from 7–21 min is ramped to 100% A. The system then runs isocratically at 100% A for 4 min (to gradient time 25 min). This is followed by a 10-min re-equilibration at 30% A and 70% B, giving a total run time of 35 min.
4. For animal tissues, where fumonisin analyses are not needed, the gradient runs from 50% A, 50% B to 100% A at 5 min followed by 5-min isocratic at 100% A. The system then returns to 50% A, 50% B for an additional 10-min re-equilibration for a total run time of 20 min.
5. The injection volume used is 20 μL. The column effluent is directly coupled to a Finnigan LTQ linear ion trap mass spectrometer (MS). The MS operates in the electrospray ionization (ESI) positive ion mode with an ion transfer capillary temperature of 210°C, and the sheath gas is nitrogen. For MS/MS of all compounds, the collision energy is 35% and the MS tuning parameters for all compounds are based on optimization for Sa 1-P.
6. The main differences between the analysis of plant tissues and those of animal tissues or mammalian cell cultures are in the compounds analyzed. For plant tissues, the instrument is routinely set to scan for FB_1, FB_2, FB_3, C_{16}-So, C_{17}-So, PSo, Sa, 1-deoxySa, Pso 1-P, and Sa 1-P.
7. For the other animal tissues/cells, all scans are the same but all FBs are removed, PSo is removed, C_{16}-So is replaced with C_{20}-So, and So and So 1-P are added. A sample chromatogram showing the retention times of all compounds in one single analysis of a standard mixture is shown in Fig. 1.

3.3.4. Quantification

1. Quantification of the compounds analyzed involves the use of both internal and external standards. For FB_1, FB_2, and FB_3, external standards are used for quantification (see Note 10). For the sphingoid bases and their respective 1-phosphates, internal standards are used for quantification.
2. To quantify Pso and Sa, the internal standard C_{16}-So is used with plant tissues and C_{20}-Sa is used to quantify So, Sa, 1-deoxySa, and Pso in animal tissues and cultured cells.

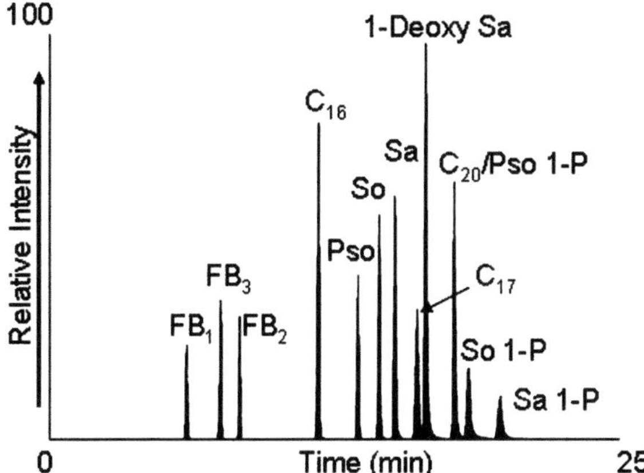

Fig. 1, Chromatogram showing the separation and relative retention times of fumonisins (FB_1, FB_2, FB_3), sphingoid bases (Pso, Sa, So, 1-deoxySa, C_{16}-So, C_{20}-Sa), and sphingoid base 1-phosphates (Pso 1-P, Sa 1-P, So 1-P, C_{17}-So 1-P). This standard was prepared as the high-level standard and contains 300 fmol/μL sphingoid bases and sphingoid base 1-phosphates, with 100 pg/μL fumonisins. Under the solvent system used, C_{20}-Sa and Pso 1-P co-elute. For additional information, see ref. (23).

3. To quantify Pso 1-P and Sa 1-P, the internal standard C_{17}-So 1-P is used. All of these compounds (C_{16}-So, C_{20}-Sa and C_{17}-So 1-P) are added to a concentration of 1 ng/μL in all plant tissue samples analyzed. In mammalian tissues or cells, the extraction buffer itself is prepared containing 60 pmol/mL C_{20}-Sa and C_{17}-So 1-P.
4. Both internal standards are stable for at least 1 week at room temperature (20–22°C). The integrated area of the standard is then compared to the area of the analyte of interest, and the ratio of areas is used to determine the quantity of the analyte by following Eq. 1, below:

$$\frac{amt\,''x''}{mg\,tissue} = \frac{amt\,IS\left(\dfrac{area\,''x''}{area\,IS}\right)}{mg\,tissue} \tag{1}$$

Variables: "*x*" = the compound of interest, IS = the internal standard

4. Notes

1. While extracellular fumonisin concentration is usually high compared to the intracellular concentration, careful rinsing of the attached cells should remove the extracellular fumonisins to a very low level. This can be controlled by experimentally

determining the residual fumonisins in cells dosed and harvested at $t=0$.

2. Other sphingoid bases and fumonisins can be detected using the described procedures but the lack of appropriate pure standards makes quantification difficult. For example, we have used this method with only slight modifications to detect methylated sphingoid bases, methylated fumonisins, fumonisins of the "C" series, and hydrolyzed fumonisins.
3. The soils can be broken up immediately upon removal from the pot by gently rolling the root mass inside a plastic bag. Gently roll the root mass and squeeze gently to break the mass apart. In this way, a large amount of soil can be dislodged prior to putting the roots into the water bath. This is useful in order to prolong the time before having to change the water due to soil accumulation. This also allows easy collection of soils in the case that analysis of soils would be useful. If soils are collected, then they must be allowed to thoroughly dry before sealing the plastic bags. Typically, this is done by leaving the open bags in a well-ventilated area and frequently mixing the soil. Drying may take several days and up to a week.
4. No amount of washing will get rid of all the soil and there is a trade off between loss of root mass and removal of all the soil. The trick is to be consistent and treat all samples with the same amount of cleaning.
5. Freeze-drying is a good opportunity to further clean dirt from the roots. As the material is dried, it is fairly easy to shake off a little more dirt, and then carefully tease some of the particles stuck in the roots out using forceps and a probe over a paper towel. The cleaned tissues can then be transferred into glass test tubes where one final cleaning step may occur prior to grinding/pulverizing. To perform this final cleaning step, first insert a sharp probe into the tissues (all inside the test tube) and agitate it briefly to allow particle movement. You should notice some dirt "powder" begin to accumulate in the bottom of the tube. Using the probe, hold as much of the tissues as possible to one side of the test tube, and gently shake out the dirt past the tissues by inverting the tube slowly and tapping on the side.
6. The rocking is gentle but ensure that the tissues remain in full contact with the extraction solvent. What you want to avoid is tissue spread (adhered) around the tube that is not in contact with the extraction solvent.
7. The preparation of Sa 1-P stock solutions is very tedious and can be tricky. Very close observation of the solutions is required to ensure that the material is actually dissolved, and

not simply finely suspended. As such, whenever stocks are stored cold, the same observations must be repeated upon warming the solutions. You should be very careful that the solutions are clear and free of particulate before continuing.

8. Other C-18 columns may be useful for the LC–MS analyses described here, but care must be taken to ensure that detection of the sphingoid base 1-phosphates is possible with the chosen column. Many columns we have tried simply do not work for the detection of the 1-phosphates. We have previously used Metachem Inertsil 5 μm ODS-3 columns 150 × 3 mm (Metachem Technologies, Inc., Torrance, CA, USA) successfully, but these columns have changed manufacturer since our use, and now work less consistently for detection of sphingoid base 1-phosphates.
9. The choice to cool the autosampler tray is a difficult one. Tray cooling is useful when your samples may degrade easily, and there is a need to keep them for further analyses. The lower temperature, however, definitely impacts the solubility of the sphingoid base 1-phosphates, which are already difficult to maintain in solution. Experimentation with this aspect may be warranted to determine optimal settings for an individual application.
10. If the use of an internal standard is required for fumonisin quantification, it is very straightforward to use ^{13}C FB_1, available from Sigma-Aldrich (St. Louis, MO, USA). When using ^{13}C FB_1 a scan needs to be added to the analysis at mass 756.3 μm. It is important to note that the response factor with LC–MS for ^{13}C-FB_1 differs from that of unlabeled FB_1 when using the LTQ linear ion trap, and this difference needs to be taken into account when quantifying. This is also true for the sphingoid bases and sphingoid base 1-phosphates when compared to the C_{16}-So, C_{20}-Sa, and C_{17}-So 1-P internal standard. When comparing the differences between treatments and controls this is usually not a problem, as the differences are extremely large. For absolute quantification, however, the relative response factors are needed to be determined and may vary with instrumentation and matrix.

References

1. Cook RJ (1981) Water relations in the biology of *Fusarium*. In: Nelson PE, Toussoun TA, Cook RJ (eds) *Fusarium*: diseases, biology, and taxonomy. The Pennsylvania State University Press, University Park, PA
2. Kommedahl T, Windels CE (1981) Root-, stalk-, and ear-infecting *Fusarium* species on corn in the USA. In: Nelson PE, Toussoun TA, Cook RJ (eds) *Fusarium*: diseases, biology, and taxonomy. The Pennsylvania State University Press, University Park, PA
3. EHC 219 (2000) Environmental health criteria 219: fumonisin B1. In: Marasas WHO, Miller JD, Riley RT, Visconti A (eds) International Programme on Chemical Safety. United Nations Environmental Programme,

The International Labour Organization, and the World Health Organization, Geneva, Switzerland, pp 1–150

4. Marasas WFO, Kellerman TS, Gelderbloom WCA, Coetzer JAW, Theil PG, Van der Lugt JJ (1988) Leukoencephalomalacia in a horse induced by fumonisin B1 isolated from *Fusarium moniliforme*. Onderstepoort J Vet Res 55:197–203
5. Ross PF, Nelson PE, Richard JL, Osweiler GD, Rice LG, Plattner RD et al (1990) Production of fumonisins by *Fusarium moniliforme* and *Fusarium proliferatum* isolates associated with equine leukoencephalomalacia and a pulmonary edema syndrome in swine. Appl Environ Microbiol 56:3225–3226
6. IARC (2002) Some traditional medicines, some mycotoxins, naphthalene and styrene (IARC Monographs on the Evaluation of Carcinogenic Risks to Humans, vol 82). IARC Press, Lyon, France, pp 275–366
7. Marasas WFO, Riley RT, Hendricks KA, Stevens VL, Sadler TW, Gelineau-van WJ et al (2004) Fumonisins disrupt sphingolipid metabolism, folate transport, and neural tube development in embryo culture and in vivo: a potential risk factor for human neural tube defects among populations consuming fumonisin-contaminated maize. J Nutr 134:711–716
8. Williams LD, Glenn AE, Zimeri AM, Bacon CW, Smith MA, Riley RT (2007) Fumonisin disruption of ceramide biosynthesis in maize roots and the effects on plant development and *Fusarium verticillioides*-induced seedling disease. J Agric Food Chem 55:2937–2946
9. Desjardins AE, Plattner RD, Nelsen TC, Leslie JF (1995) Genetic analysis of fumonisin production and virulence of *Gibberella fujikuroi* mating population A (*Fusarium moniliforme*) on maize (*Zea mays*) seedlings. Appl Environ Microbiol 61:79–86
10. Rheeder JP, Marasas WFO, Vismer HF (2002) Production of fumonisin analogs by *Fusarium* species. Appl Environ Microbiol 68: 2101–2105
11. Bartok T, Szecsi A, Szekeres A, Mesterhazy A, Bartok M (2006) Detection of new fumonisin mycotoxins and fumonisin-like compounds by reversed-phase high-performance liquid chromatography/electrospray ionization ion trap mass spectrometry. Rapid Commun Mass Spectrom 20:2447–2462
12. Wang E, Norred WP, Bacon CW, Riley RT, Merrill AH Jr (1991) Inhibition of sphingolipid biosynthesis by fumonisins. Implications for diseases associated with *Fusarium moniliforme*. J Biol Chem 266:14486–14490
13. Riley RT, Enongene E, Voss KA, Norred WP, Meredith FI, Sharma RP et al (2001) Sphingolipid perturbations as mechanisms for fumonisin carcinogenesis. Environ Health Perspect 109:301–308
14. Abbas HK, Tanaka T, Duke SO, Porter JK, Wray EM, Hodges L et al (1994) Fumonisin- and AAL-toxin-induced disruption of sphingolipid metabolism with accumulation of free sphingoid bases. Plant Physiol 106:1085–1093
15. Merrill AH, Jr SMC, Wang E, Voss KA, Riley RT (2001) Sphingolipid metabolism: roles in signal transduction and disruption by Fumonisins. Environ Health Perspect 109: 283–289
16. Maceyka M, Milstien S, Spiegel S (2009) Sphingosine-1-phosphate: the Swiss army knife of sphingolipid signaling. J Lipid Res 50:S272–S276
17. Ng CK-Y, Hetherington AM (2001) Sphingolipid-mediated signalling in plants. Ann Bot 88:957–965
18. Worrall D, Ng CK-Y, Hetherington AM (2003) Sphingolipids, new players in plant signaling. Trends Plant Sci 8:317–320
19. Lynch DV, Dunn TM (2004) An introduction to plant sphingolipids and a review of recent advances in understanding their metabolism and function. New Phytol 61:677–702
20. Coursol S, Stunff HE, Lynch DV, Gilroy S, Assman SM, Spiegel S (2005) *Arabidopsis* sphingosine kinase and the effects of phytosphingosine-1-phosphate on stomatal aperture. Plant Physiol 137:724–737
21. Shi L, Bielawski J, Mu J, Dong H, Teng C, Zhang J et al (2007) Involvement of sphingoid bases in mediating reactive oxygen intermediate production and programmed cell death in *Arabidopsis*. Cell Res 17:1030–1140
22. Glenn AE, Zitomer NC, Zimeri AM, Williams LD, Riley RT, Proctor RH (2008) Transformation-mediated complementation of a FUM gene cluster deletion in *Fusarium verticillioides* restores both fumonisin production and pathogenicity on maize seedlings. Mol Plant Microbe Interact 21:87–97
23. Zitomer NC, Glenn AE, Bacon CW, Riley RT (2008) A single extraction method for the analysis by liquid chromatography/tandem mass spectrometry of fumonisins and biomarkers of disrupted sphingolipid metabolism in tissues of maize seedlings. Anal Bioanal Chem 391:2257–2263
24. Riley RT, Voss KA (2006) Differential sensitivity of rat kidney and liver to fumonisin toxicity: organ-specific differences in toxin

accumulation and sphingoid base metabolism. Toxicol Sci 92:335–345

25. Enongene EN, Sharma RP, Bhandari N, Miller JD, Meredith FI, Voss KA et al (2002) Persistence and reversibility of the elevation in free sphingoid bases induced by fumonisin inhibition of ceramide synthase. Toxicol Sci 67:173–181

26. Zitomer NC, Mitchell T, Voss KA, Bondy GS, Pruett ST, Garnier-Amblard EC et al (2009) Ceramide synthase inhibition by fumonisin B_1 causes accumulation of 1-deoxysphinganine: a novel category of bioactive 1-deoxysphingoid bases and 1-deoxyhydroceramides biosynthesized by mammalian cell lines and animals. J Biol Chem 284:4786–4795

Chapter 16

Determination of Fumonisins B_1 and B_2 in Maize Food Products by a New Analytical Method Based on High-Performance Liquid Chromatography and Fluorimetric Detection with Post-column Derivatization

Marilena Muscarella, Sonia Lo Magro, Donatella Nardiello, Carmen Palermo, and Diego Centonze

Abstract

A sensitive and selective analytical method for the quantitative determination of fumonisins B_1 (FB_1) and B_2 (FB_2) in maize-based foods for direct human consumption is described. The method, based on high-performance liquid chromatography and fluorescence detection, presents a rapid and automated online post-column derivatization, performed with *o*-phthalaldehyde and *N*,*N*-dimethyl-2-mercaptoethylamine (Thiofluor™). A complete separation of fumonisins is achieved in less than 13 min by using a C18 column and a gradient elution. Fumonisins are extracted from the sample with a mixture of water, acetonitrile, and methanol. The filtered extract is purified by immunoaffinity column and FB_1 and FB_2 are eluted with methanol. The method has been successfully validated, and performances comply with criteria of the Regulation EC No 401/2006.

Key words: Fumonisins, FB_1, FB_2, Liquid chromatography, Fluorescence detection, Post-column chemical derivatization, *o*-phthalaldehyde, *N*,*N*-dimethyl-2-mercaptoethylamine (Thiofluor™)

1. Introduction

Fumonisins are a group of naturally occurring mycotoxins produced by several fungal species like *Fusarium moniliforme* and *Fusarium proliferatum*, and pathogens of maize and other cereals, such as sorghum and rice (1, 2). Among fumonisins, only for the sum of FB_1 and FB_2, the European Union has set maximum levels in a variety of foods (maize-based products, snacks and cereals for breakfast, baby foods, etc.) intended for direct human

Otto Holst (ed.), *Microbial Toxins: Methods and Protocols*, Methods in Molecular Biology, vol. 739,
DOI 10.1007/978-1-61779-102-4_16, © Springer Science+Business Media, LLC 2011

consumption (3). Therefore, the accurate determination of fumonisins is a topic of increasing interest and the development of sensitive and reliable methods represents an important aspect for food assurance and quality control. In the past decades, a variety of analytical methods for determination of fumonisins B_1 and B_2, including enzyme-linked immunosorbent assay (ELISA) (4), gas chromatography (5), liquid chromatography coupled to mass spectrometry (6, 7), and capillary electrophoresis (8) have been reviewed (9). Among these, chromatographic determination based on reversed-phase high-performance liquid chromatography (RP-HPLC) and fluorescence detection (FLD) with *o*-phthalaldehyde (OPA) post-chemical derivatization, combined to a minimal sample preparation and cleanup (10) assures a fast and accurate analytical response in real time, which is very useful in plans of monitoring and in the official control analyses.

2. Materials

2.1. Sample Extraction and Cleanup

1. Extraction solvent (MeCN:MeOH:H_2O, 30:30:40, by vol). Mix 30 volume parts of acetonitrile with 30 volume parts of methanol and 40 volume parts of water. Water and organic solvents should be of HPLC analytical grade (see Note 1).
2. Phosphate buffered saline (PBS). Dissolve 8.0 g of NaCl, 0.2 g of KCl, 3.026 g of $Na_2HPO_4 \cdot 12H_2O$, and 0.2 g of KH_2PO_4 in approximately 990 mL of water. Adjust pH to 7.0 with concentrated HCl and add water to 1 L. All the reagents are of ACS grade.
3. Immunoaffinity column (IAC). IAC columns (FUMONIPREP®, R-Biopharm Rhône, Ltd.) should have a total capacity of not less than 5 μg of fumonisins. They should be stored at 2–8°C and warmed up to room temperature (25°C) before use. Do not freeze. The antibody of the IAC can be denaturated by extreme conditions of temperature and pH.

2.2. Standard Stock and Working Solutions

1. Fumonisin stock solutions (1,000 mg/L). Prepare a stock solution of FB_1 and a stock solution of FB_2 in methanol at a mass concentration of 1,000 mg/L, by dissolving 1 mg of each fumonisin (Sigma–Aldrich, Steinheim, Germany) in 1 mL of MeOH. Store the solutions at –20°C. Fumonisin stock solutions are stable for at least 6 months (see Note 2).
2. Fumonisin stock solution (10 mg/L). Prepare a stock solution by pipetting into a 10-mL volumetric flask, 100 μL of each fumonisin stock solution at concentration of 1,000 mg/L. Dilute to the mark with MeOH. This solution is stable for at least 6 months when stored at –20°C.

3. Fumonisin standard solution for HPLC. Prepare a fumonisin standard solution at a concentration of 640 μg/L for each fumonisin by pipetting into a 2-mL vial, 64 μL of the 10 mg/L fumonisin standard solution, and 936 μL of mobile phase at the initial gradient elution conditions (0.1 *M* phosphate buffer, pH 3.15/MeOH, 40:60, v/v). From the 640 μg/L solution, prepare 4 fumonisin standard solutions in 2-mL vials according to Table 1, and add mobile phase (phosphate buffer 0.1 *M*, pH 3.15/MeOH, 40:60, v/v) to 1 mL.

2.3. Experimental Conditions of Chromatographic Separation and FLD

1. HPLC mobile phase. Prepare solvent A composed of sodium dihydrogen phosphate solution. Dissolve 13.799 g of $NaH_2PO_4 \times H_2O$ (ACS grade) in about 990 mL of water. Adjust to pH 3.15 with concentrated *o*-phosphoric acid and add water to 1 L. Filter the solution through a membrane filter (Whatman, 0.45 μm). Solvent B is MeOH. The gradient elution conditions are listed in Table 2.

Table 1
Preparation of fumonisin standard solutions for HPLC

Fumonisin standard solution for HPLC	Volume (μL) of the 640 μg/L fumonisin solution	Volume (μL) of mobile phase-solvent A added	Final fumonisin concentration (μg/L)	
			FB_1	FB_2
1	750	250	480	480
2	500	500	320	320
3	375	625	240	240
4	250	750	160	160

Table 2
Gradient elution program. (A): 0.1 M phosphate buffer, pH 3.15, and (B): MeOH

Step N.	Time (min)	% A	% B
1	0	40	60
2	5	35	65
3	8	25	75
4	10	40	60
5	15	40	60

2. Derivatization solution. The daily prepared derivatization solution is composed of OPA and *N,N*-dimethyl-2-mercaptoethylamine (Thiofluor™) in potassium borate buffer (*o*-phthalaldehyde diluent, OD104, supplied by Pickering Laboratories – Mountain View, USA). Dissolve 0.2 g of Thiofluor™ in about 1 mL of OD104, add OPA prepared by dissolving 0.2 g in about 1 mL of methanol, and dilute with OD104 to a final volume of 200 mL. The reagent should be kept in the dark in an amber bottle.

2.4. Equipment

1. HPLC system. The chromatographic system consists of the following:
   (a) Binary pump equipped with a micro vacuum degasser
   (b) Thermostated autosampler
   (c) Column compartment
   (d) Fluorescence detector, fitted with a flow cell and set at 343 nm (excitation) and 445 nm (emission).
   The system is interfaced to a personal computer, and a chromatographic software is used to control instruments and for data acquisition and processing.
2. Analytical reverse-phase separating column. ZORBAX Eclipse® XDB-C18 (150 × 4.6 mm i.d., particle size 5 μm) from Agilent Technologies kept at 40°C. Flow rate: 0.8 mL/min. Injection volume: 100 μL.
3. Online post-column chemical derivatization system. Commercially available system operating at a flow rate of 0.4 mL/min, supplied by LabService Analytica (Bologna, Italy) and consisting of the following:
   (a) Double-piston pump (model K-120)
   (b) Thermostatable post-column reactor (model CRX 400) set at 40°C, equipped with a 0.5 mL knitted reaction coil.

3. Methods

3.1. Sample Extraction

1. Weigh 5 g of homogenized sample into a 50-mL polypropylene centrifuge tube and add 12.5 mL of extraction solvent (MeCN:MeOH:H_2O, 30:30:40, by vol.).
2. Shake for 20 min with orbital shaker, and then sonicate for other 20 min.
3. Centrifuge at 2,112 × *g* for 10 min at 25°C. Remove and store the supernatant in a 50-mL flask.
4. Extract again the remaining solid material by adding 12.5 mL of the extraction solvent into the polypropylene centrifuge tube. Repeat the shaking, sonication, and centrifugation processes.

5. Mix the two supernatants and filter through paper filter, avoiding the transfer of solid material.
6. Pipet 3 mL of the filtered extract into a flask, add 12 mL of PBS and mix thoroughly; 10 mL of this extract is used for IAC cleanup.

3.2. IAC Cleanup

1. Remove the cap from the column and discard the storage buffer.
2. Pipet 10 mL of the filtered extract diluted in PBS. Let the extract flow through the column at approximately 1–2 drops per second, and discard the eluate; a weak and steady pressure is essential to allow the fumonisin binding to the antibody.
3. Wash the column with 10 mL of PBS at a flow rate lower than 0.5 mL/min and discard the eluate.
4. Elute fumonisins with 4 mL of MeOH at a flow rate not more than 1 drop per second or by gravity. Push air through the column to collect the last few drops of eluate.
5. Evaporate the eluate to dryness under a stream of nitrogen at 40°C.
6. Dissolve the dried residue with 500 μL of mobile phase-solvent A-step 1 (0.1 *M* phosphate buffer, pH 3.15/MeOH, 40:60, v/v) and filter through inorganic membrane Anotop 10 LC (0.2 μm, Whatman).

3.3. HPLC–FLD Determination

In the analysis of a new matrix, the method selectivity should be checked. For this purpose, the comparison between blank (see Note 3) and spiked samples chromatograms is essential in order to verify the absence of interfering peaks in the retention time-window of interest (±2.5% of the retention time of each fumonisin).

3.4. Detection and Quantification of Fumonisins

1. Identify fumonisin B_1 and B_2 by comparing the retention times of each real sample peak with those of a fumonisin standard solution. As an example, in Fig. 1 a typical chromatogram of a fumonisin standard solution and a cornmeal sample is shown.
2. The calibration curve for FB_1 and FB_2 is obtained by three series of analyses on three different days of standard solutions at a concentration of 160, 240, 320, 480, and 640 μg/L, prepared as described in Table 1. A good linearity for both fumonisins, with correlation coefficients higher than 0.9997, is expected.
3. Detection (LOD) and quantification (LOQ) limits are calculated according to the equations: $LOD = 3.3 \cdot s_a/b$ and $LOQ = 10 \cdot s_a/b$, where s_a is the standard deviation of the intercept and b is the slope of the regression line obtained from the calibration curves. The standard deviations of slope

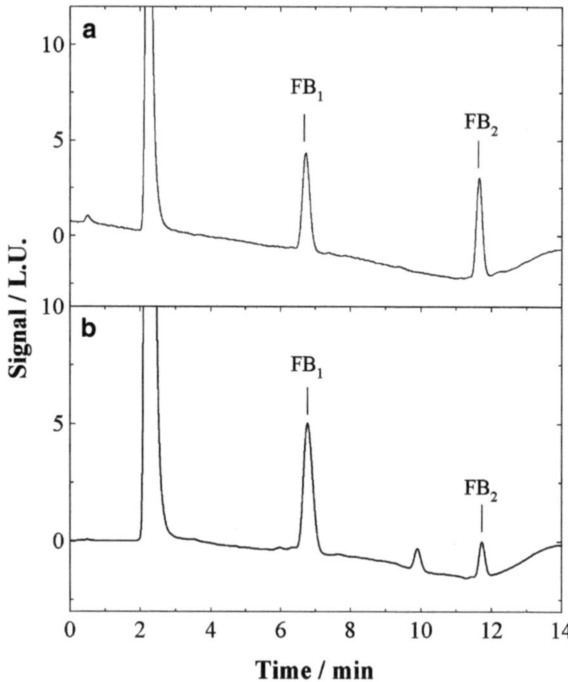

Fig. 1. Typical chromatograms of a standard solution of fumonisins B_1 and B_2 at a concentration of 160 μg/L each (**a**) and a cornmeal sample (**b**).

and intercept are estimated at the 95% confidence level. Limits of detection of 4 and 5 μg/L (corresponding to 5 and 6 μg/kg in matrix) have been achieved for FB_1 and FB_2. Typical values of LOQ around 15 μg/L (corresponding to 19 μg/kg in matrix) are expected. Calibration parameters should be checked at the beginning of each working session.

4. Quantify the concentrations (reported in μg/L) of the fumonisins in the eluate by comparing sample peak areas to the calibration curve. Calculate the corresponding concentration in the sample (μg/kg) by multiplying by the dilution factor 1.25, evaluated for FB_1 and FB_2 taking into account the sample extraction and cleanup.

3.5. Recovery and Precision

1. The method should be periodically tested for intra- and inter-day assay reproducibility to determine both the trueness and precision. Experimental values of recovery and relative standard deviations should be in agreement with provisions of Regulation EC No 401/2006, dealing with the official control analyses of mycotoxins in foodstuffs.
2. Precision and recovery data are obtained from the analysis of samples fortified, prior to the extraction, with proper amounts of a standard solution of FB_1 and FB_2 at final concentrations

Table 3
Within-laboratory reproducibility and recovery evaluated for spiked cornflakes samples

Fumonisin	Contamination level (μg/kg)	Recovery % (mean value ± SD)	RSD (%) Intra-day (n=6)	RSD (%) Inter-day (n=12)
B_1	100	94.2 ± 6.8	3.4	7.2
	200	89.3 ± 6.9	3.7	7.7
	300	87.0 ± 2.4	1.9	2.7
	400	76.5 ± 8.4	4.1	11
B_2	100	71.4 ± 6.5	4.7	9.1
	200	74.8 ± 6.2	4.5	8.3
	300	69.9 ± 3.3	2.3	4.6
	400	66.8 ± 4.9	3.6	7.0

of 0.5, 1, and 1.5 times the maximum limit (ML) established in Regulation EC No 1126/2007.

3. Analyses are performed in different days with the same instruments, but with different operators and instrumental calibrations.
4. Recoveries of fumonisins are evaluated by comparing the concentration of spiked samples determined by interpolation on the calibration curve with the nominal concentration of the fortification level.
5. Typical precision and recovery data evaluated for each fumonisin in spiked cornflake samples at four fortification levels (100, 200, 300, and 400 μg/kg) are summarized in Table 3. The values derived from the method validation procedure performed according to the European regulation (11, 12).

4. Notes

1. All the solutions should be prepared in water that has a resistivity of 18.2 MΩ-cm and total organic content of less than five parts per billion.
2. Fumonisins are very hazardous substances. FB_1 has been classified by the International Agency for Research on Cancer (IARC) as a probable human carcinogen (class 2B). Observe appropriate safety precautions for handling fumonisins. Suitable protective clothing, including rubber gloves, safety glasses, and laboratory coats should be used throughout

the analysis. All materials, glassware, and toxin-containing solutions should be decontaminated before discarding or reusing by soaking for at least 30 min in a 5% solution of sodium hypochlorite.

3. Blank samples can be tested for the absence of fumonisins by using common ELISA screening methods.

Acknowledgments

The authors gratefully acknowledge P. D'Antini (Istituto Zooprofilattico Sperimentale della Puglia e della Basilicata) for the technical assistance. Ministero della Salute (Rome, Italy) is thanked for providing the financial support.

References

1. Gelderblom WCA, Jaskiewicz K, Marasas WFO, Thiel PG, Horak MJ, Vleggaar R (1988) Fumonisins-novel-mycotoxins with cancer promoting activity produced by *Fusarium moniliforme*. Appl Environ Microbiol 54:1806–1811
2. Trucksess MW (2001) Mycotoxins. J AOAC Int 84:202–211
3. European Commission. Regulation (EC) No 1126/2007. Off J Eur Union L255:14–17
4. Castells M, Marín S, Sanchis V, Ramos AJ (2008) Distribution of fumonisins and aflatoxins in corn fractions during industrial cornflake processing. Int J Food Microbiol 123:81–87
5. Plattner RD, Branham BE (1994) Labelled fumonisins: production and use of fumonisin B_1 containing stable isotopes. J AOAC Int 77:525–532
6. Paepens C, De Saeger S, Van Poucke C, Dumoulin F, Van Calenbergh S, Van Peteghem C (2005) Development of a liquid chromatography/tandem mass spectrometry method for the quantification of fumonisin B_1, B_2 and B_3 in cornflakes. Rapid Commun Mass Spectrom 19:2021–2029
7. Zöllner P, Mayer-Helm B (2006) Trace mycotoxin analysis in complex biological and food matrices by liquid chromatography-atmospheric pressure ionisation mass spectrometry. J Chromatogr A 1136:123–169
8. García-Cãnas V, Cifuentes A (2007) Detection of microbial food contaminants and their products by capillary electromigration techniques. Electrophoresis 28:4013–4030
9. Shephard GS (1998) Chromatographic determination of the fumonisins mycotoxins. J Chromatogr A 815:31–39
10. Muscarella M, Lo Magro S, Nardiello D, Palermo C, Centonze D (2008) Development of a new analytical method for the determination of fumonisins B_1 and B_2 in food products based on high performance liquid chromatography and fluorimetric detection with post-column derivatization. J Chromatogr A 1203: 88–93
11. European Commission. Regulation (EC) No 882/2004. Off J Eur Union L165:1–141
12. European Commission. Regulation (EC) No 401/2006. Off J Eur Union L70:12–34

Chapter 17

A Confirmatory Method for Aflatoxin M_1 Determination in Milk Based on Immunoaffinity Cleanup and High-Performance Liquid Chromatography with Fluorometric Detection

Marilena Muscarella, Sonia Lo Magro, Carmen Palermo, and Diego Centonze

Abstract

A sensitive and reliable analytical method based on immunoaffinity chromatography cleanup followed by HPLC separation and fluorimetric detection is described for the quantitative determination of aflatoxin M_1 in milk. The chromatographic separation is accomplished by using a C18 column and a gradient elution with methanol, acetonitrile, and water. No extraction solvent process is required and a minimal milk sample cleanup is performed by a direct loading of the immunoaffinity columns and elution with methanol. The method has been successfully validated according to Decision EC No 657/2002 by using the conventional validation approach. The results of the validation process demonstrate the agreement of the method with the provisions of Regulation EC No 401/2006.

Key words: Aflatoxin M_1, Milk, Liquid chromatography (HPLC), Fluorescence detection, Immunoaffinity cleanup

1. Introduction

Aflatoxins are secondary metabolites produced by molds belong ing to *Aspergillus* species (*Aspergillus flavus* and *Aspergillus parasiticus*), which can be found in a wide variety of agricultural products and animal fodder (1, 2) as a result of molds contamination before or during harvest or improper storage (3). Aflatoxin M_1 (AFM_1) is metabolized and accumulated in milk in animals eating aflatoxin B_1 contaminated foodstuffs (4). AFM_1 contamination represents a risk for human health, owing to the hepatoxic

Otto Holst (ed.), *Microbial Toxins: Methods and Protocols*, Methods in Molecular Biology, vol. 739,
DOI 10.1007/978-1-61779-102-4_17, © Springer Science+Business Media, LLC 2011

and carcinogenic activity displayed, and therefore it has been included in the IARC group 1 (5). In the last years, for the determination of AFM_1 in milk samples several procedures have been developed, including immunological assay ELISA (6), electrochemical immunosensors systems (7), and methods based on high-performance liquid chromatography (8, 9). The quantitative determination of AFM_1 in confirmatory routine analyses of official samples is usually performed by immunoaffinity column (IAC) sample cleanup, followed by normal or reversed-phase HPLC separation with fluorimetric detection (10) because of characteristics of specificity, high sensitivity, and simplicity of operation.

2. Materials

2.1. Sample Cleanup

1. IAC: The column (RIDA® Aflatoxin, R-Biopharm AG) contains monoclonal antibodies conjugated to Sepharose raised against aflatoxins B_1, B_2, G_1, G_2, and M_1. The column should have a capacity of approximately 40 ng aflatoxin. The column should be stored at 2–8°C and warmed up to room temperature (25°C) before use. Do not freeze. The antibody included in the IAC can be denaturated by extreme temperature and pH. Excessive pressure on the column or extreme vacuum in the vacuum unit may cause compression of the gel and consequently low recovery.

2.2. Standard Stock and Working Solutions

1. AFM_1 stock solution (200 μg/L). Prepare a stock solution of AFM_1 in methanol (see Note 1) at a concentration of 200 μg/L from a certificated standard solution of AFM_1 (Supelco, Bellefonte, USA) in acetonitrile of about 10 mg/L (see Note 2). Pipet into a 10-mL volumetric flask a proper volume (~200 μL) in dependence on the exact title of the certified material. Store the solution at –20°C in the dark. Under the described conditions, this solution is stable for 3 months (see Note 3).
2. AFM_1 stock solution (20 μg/L). Prepare an intermediate 20 μg/L stock solution by pipetting 1 mL of the 200 μg/L stock solution into a 10-mL volumetric flask. Dilute to the mark with mobile phase at the initial gradient elution conditions (methanol/acetonitrile/water, 21/24/55, by vol.), prepared into a 100-mL flask, by mixing 21 mL of methanol, 24 mL of acetonitrile, and 55 mL of water. This solution is stable for 1 month when stored at –20°C (see Note 2).
3. AFM_1 standard solution for HPLC. From a 20 μg/L AFM_1 intermediate stock solution, prepare five working standard

solutions into 2-mL amber vials, according to Table 1, and add mobile phase (methanol/acetonitrile/water, 21/24/55, by vol.) to 1 mL.

2.3. Experimental Conditions of the Chromatographic Separation and Fluorescence Detection

1. HPLC mobile phase. Chromatographic separations are performed in gradient mode with a mobile phase consisting of methanol (A), acetonitrile (B), and water (C) (see Note 1). The gradient elution conditions are reported in Table 2.

2.4. Equipment

1. HPLC system. The chromatographic system consists of the following:
   (a) Binary pump equipped with a micro vacuum degasser
   (b) Thermostated autosampler
   (c) Column compartment

Table 1
Preparation of standard solutions for HPLC

Standard solution for HPLC	Volume (μL) from a 20 μg/L solution	Volume (μL) of mobile phase added	Final AFM_1 concentration (μg/L)
1	100	900	2.0
2	75	925	1.5
3	45	955	0.9
4	30	970	0.6
5	15	985	0.3

Table 2
Gradient elution program. (A): MeOH, (B): MeCN, and (C): H_2O

Step N.	Time (min)	% A	% B	% C
1	0	21	24	55
2	6	21	24	55
3	7	45	40	15
4	10	45	40	15
5	11	21	24	55
6	17	21	24	55

(d) Fluorescence detector, fitted with a flow cell and set at 360 nm (excitation) and 440 nm (emission).
The system is interfaced, via a network chromatographic software, to a personal computer for the control of instruments, data acquisition, and processing.

2. Analytical reverse-phase separating column. ZORBAX Eclipse® XDB-C18 (250×4.6 mm i.d., particle size 5 μm) from Agilent Technologies kept at 30°C. Flow rate: 0.8 mL/min. Injection volume: 20 μL.

3. Methods

3.1. Sample Cleanup by IACs

1. Prior to sample cleanup, centrifuge milk samples at 2,112 ×*g* for 10 min at 10°C, and then weigh 10.00 ± 0.01 g of skimmed milk.
2. Before use, remove caps on top and tip from the IAC, remove the storage buffer, and then rinse the column with 2 mL of deionized water.
3. Fill the column with 1 mL of skimmed milk and connect a suitable adapter on top of the column to use a syringe as sample reservoir. Fill the syringe with the residual skimmed milk.
4. Inject the residual skimmed milk sample slowly and continuously through the column (flow rate approximately 1 drop per second) and discard the eluate.
5. Wash the column with 10 mL of deionized water. Press some air through the column for approximately 10 s to make sure that all the residual liquids will be removed from the column.
6. Remove the syringe and place a clean vial below the column in order to collect the eluate.
7. Elute AFM_1 with 0.75 mL of MeOH at a flow rate not more than 1 drop per second or by gravity. Push air through the column to collect the eluate entirely.
8. Evaporate the eluate to dryness under a stream of nitrogen at 45°C.
9. Dissolve the dried residue with 500 μL of mobile phase (methanol/acetonitrile/water, 21:24:55, by vol.) and filter through inorganic membrane Anotop 10 LC (0.2 μm, Whatman). The methanolic eluates can be stored for a week at −20°C, protected against light (see Note 2).

3.2. HPLC–FLD Determination

In the analysis of milk from different animal species, the method selectivity should be checked. For this purpose, the comparison among chromatograms of blank (see Note 4) and spiked samples is essential in order to verify the absence of interfering peaks in the retention time-window of interest (±2.5% of the retention time of AFM_1).

3.3. Detection and Quantification of AFM_1

1. The identification of AFM_1 is performed by comparing the retention time of each real sample peak with that of a standard solution. As an example, in Fig. 1 a typical chromatogram of an AFM_1 standard solution and a spiked bovine milk sample are shown.
2. Calibration curves for AFM_1 can be obtained by three series of analyses on three different days, injecting AFM_1 working standard solutions at a concentration of 0.3, 0.6, 0.9, 1.5, and 2.0 μg/L, prepared as described in Table 1. A good linearity with correlation coefficient higher than 0.9995 is expected.
3. Detection (LOD) and quantification (LOQ) limits are calculated according to the equations $LOD = 3.3 \cdot s_a/b$ and

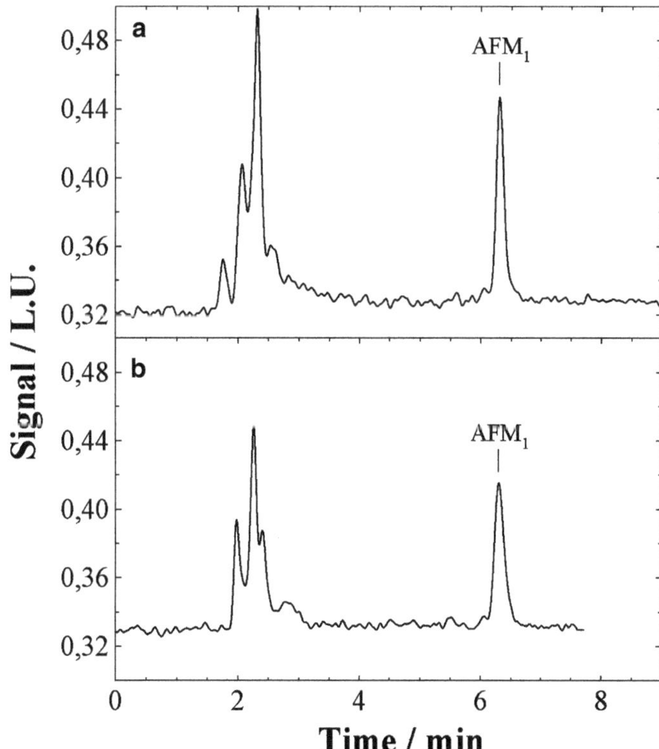

Fig. 1. Typical chromatograms of AFM_1 at a concentration of 1.5 μg/kg (**a**), and a spiked bovine milk sample at 0.075 μg/kg of AFM_1 (**b**).

LOQ = 10 · s_a/b, where s_a is the standard deviation of the intercept and b is the slope of the regression line obtained from the calibration curves. The standard deviations of slope and intercept are estimated at the 95% confidence level. Limits of detection and quantification of 0.12 and 0.30 μg/L, respectively (corresponding to 0.006 and 0.015 μg/kg in matrix) are expected. Calibration parameters should be checked at the beginning of each working session.

4. Quantify the aflatoxin concentration (reported in μg/L) in the eluate by comparing sample peak area to the calibration curve. Calculate the corresponding concentration in the sample (μg/kg) by multiplying by the dilution factor 0.05, evaluated for AFM_1 taking into account the sample cleanup.

3.4. Recovery and Precision

1. The method should be periodically tested to determine its trueness and intra- and inter-day precision. Experimental values of recovery and relative standard deviations should be in agreement with provisions of Regulation EC No 401/2006, dealing with the official control analyses of mycotoxins in foodstuffs (11).
2. Precision and recovery data are obtained from the analysis of samples fortified, prior to the extraction, with proper amounts of AFM_1 at final concentrations of 0.5, 1, and 1.5 times the maximum limit (ML) established in Regulation EC No 1881/2006 (12).
3. Analyses should be performed in different days with the same instruments, but with different operators and instrumental calibrations.
4. Recovery values are evaluated by comparing the concentration of spiked samples, determined by interpolation on the calibration curve, with the nominal fortification level.
5. In Table 3 typical precision and recovery data are summarized, evaluated for AFM_1 in spiked bovine milk samples at

Table 3
Within-laboratory reproducibility and recovery evaluated for spiked bovine milk samples

Contamination level (μg/kg)	Recovery % (mean value ± SD)	RSD (%) Intra-day (n = 6)	RSD (%) Inter-day (n = 18)
0.025	92 ± 14	12	15
0.050	90.4 ± 9.4	9.4	10
0.075	91.4 ± 7.3	7.9	8.4

fortification levels of 0.025, 0.050, and 0.075 μg/kg. The values derived from the method validation procedure (10) performed according to the European regulation (11, 13).

4. Notes

1. During the analysis, reagents of analytical grade should be used. All the solutions should be prepared in water that has a resistivity of 18.2 MΩ-cm. Water and organic solvents used as eluents should be for HPLC analysis.
2. AFM_1 is light-sensitive; therefore, avoid exposure of the standard solutions and sample extracts to direct light. Original samples should be kept refrigerated, dry, protected against light, and well sealed.
3. Particular care should be taken as aflatoxins are toxic and carcinogenic substances. Avoid contact of standard solutions and sample extracts with the skin (use gloves). Decontamination of materials, glassware, and toxin-containing solutions should be carried out by using a sodium hypochlorite (bleach) solution (10% v/v) for about 18 h before reusing or discarding.
4. Blank samples can be tested for the absence of AFM_1 by using common ELISA screening methods.

Acknowledgments

The authors gratefully acknowledge P. D'Antini (Istituto Zooprofilattico Sperimentale della Puglia e della Basilicata) for the technical assistance. This work was supported by Ministero della Salute (Rome, Italy).

References

1. Peraica M, Radic B, Lucic A, Pavlovic M (1999) Toxic effects of mycotoxins in human. B World Health Organ 77:754–766
2. Eaton DL, Groopman JD (1994) The toxicology of aflatoxins. Academic Press, New York
3. Sweeney MJ, Dobson ADW (1998) Mycotoxin production by *Aspergillus*, *Fusarium* and *Penicillium* species. Int J Food Microbiol 43:141–158
4. van Egmond HP (1989) Mycotoxins in dairy products. Elsevier Applied Science, London, NY, pp 1–54
5. International Agency for Research on Cancer (IARC). (2002) *IARC Monographs on the Evaluation of Carcinogenic Risks to Humans* 56:171–176
6. Magliulo M, Mirasoli M, Simoni P, Lelli R, Portanti O, Roda A (2005) Development and validation of an ultrasensitive chemiluminescent enzyme immunoassay for aflatoxin M1 in milk. J Agric Food Chem 53:3300–3305
7. Badea M, Micheli L, Messia MC, Cardigliota T, Marconi E, Mottram T, Valesco-Garcia M, Moscone D, Palleschi G (2004) Aflatoxin M1

determination in raw milk using a flow-injection immunoassay system. Anal Chim Acta 520:141–148

8. Gilbert J, Anklam E (2002) Validation of analytical methods for determining mycotoxins in foodstuffs. TrAC – Trend Anal Chem 21:468–486
9. Cavaliere C, Foglia P, Pastorini E, Samperi R, Laganà A (2006) Aflatoxin M1 determination in cheese by liquid chromatography-tandem mass spectrometry. J Chromatogr A 1101:69–78
10. Muscarella M, Lo MS, Palermo C, Centonze D (2007) Validation according to European Commission Decision 2002/657/EC of a confirmatory method for aflatoxin M1 in milk based on immunoaffinity columns and high performance liquid chromatography with fluorescence detection. Anal Chim Acta 594:257–264
11. European Commission. Regulation (EC) No 401/2006. Off J Eur Union L 70: 12–34
12. European Commission. Regulation (EC) No 1881/2006. Off J Eur Union L 364:5–24
13. European Commission. Regulation (EC) No 882/2004. Off J Eur Union L 165:1–141

Chapter 18

Simultaneous Determination of Aflatoxins B_1, B_2, G_1, and G_2 in Foods and Feed Materials

Marilena Muscarella, Marco Iammarino, Donatella Nardiello, Sonia Lo Magro, Carmen Palermo, and Diego Centonze

Abstract

A high-performance liquid chromatographic method with on-line post-column photochemical derivatization and fluorimetric detection for the simultaneous separation and quantitative determination of aflatoxin (AF) B_1, B_2, G_1, and G_2 in foodstuffs and feed materials is reported.

The chromatographic separation is accomplished by using a C_{18} column eluted with an isocratic mobile phase consisting of water, methanol, and acetonitrile. The sample preparation requires a simple extraction of aflatoxins with a mixture of water and methanol, and a purification step by immunoaffinity column clean-up. The total analysis time, including sample preparation and chromatographic separation, does not exceed 40 min with a run time of 10 min. The procedure for the determination of aflatoxins in food samples and cereals for animal consumption has been extensively validated, in agreement with Regulation (EC) No. 882/2004, demonstrating the conformity of the method with provisions of Regulation (EC) No. 401/2006 in terms of sensitivity, linearity, selectivity, and precision.

Key words: Aflatoxins B_1, B_2, G_1 and G_2, Liquid chromatography, Photochemical derivatization, Fluorescence detection

1. Introduction

Aflatoxins B_1, B_2, G_1, and G_2 (AFB_1, AFB_2, AFG_1, and AFG_2), secondary metabolites of mold fungi *Aspergillus flavus* and *Aspergillus parasiticus*, are a group of structurally related mycotoxins, which exhibit acute and chronic toxicity including mutagenic, carcinogenic, and teratogenic effects in several organisms (1–3). Due to the significant health risks associated with the presence of aflatoxins in foods and to satisfy the rigorous legal requirements, efficient techniques for the detection of aflatoxins in foods and animal feeds are required. Several procedures

Otto Holst (ed.), *Microbial Toxins: Methods and Protocols*, Methods in Molecular Biology, vol. 739,
DOI 10.1007/978-1-61779-102-4_18,

based on enzyme-linked immunosorbent assays (ELISAs) (4), and chromatographic techniques coupled with chemical pre- or post-column derivatization and fluorimetric detection have been proposed (5–7). Although ELISA methods are extensively used for rapid qualitative screenings of aflatoxins, they are not useful in providing a definitive confirmation of the toxins and an accurate quantitative determination. The present trend is the use of high-performance liquid chromatography (HPLC) because of the characteristics of specificity, high sensitivity, and simplicity of operation. The procedure for the determination of aflatoxins B_1, B_2, G_1, and G_2 in food products and materials for feed (8), based on immunoaffinity column sample clean-up and liquid chromatography coupled to photochemical derivatization and fluorescence detection, is described. Through a validation procedure, linearity, selectivity, recovery, precision, detection and quantification limits (LODs, LOQs), ruggedness of the method, and measurement uncertainty have been evaluated, demonstrating the suitability of the method in the official control of aflatoxins in foodstuffs.

2. Materials

2.1. Sample Extraction and Clean-Up

1. Extraction solvent (MeOH/H_2O 80/20, v/v). Prepare a solution of MeOH/H_2O (80/20, v/v) in a 100-mL volumetric flask by mixing 80 volume parts of methanol with 20 volume parts of water (see Note 1).
2. Phosphate-buffered saline (PBS). Dissolve 9.0 g of NaCl, 2.85 g of Na_2HPO_4 $2H_2O$, and 0.55 g of NaH_2PO_4 H_2O in approximately 990 mL of water. Add 1.0 mL of Tween 20®. Adjust pH to 7.2 with concentrated HCl or 1 M NaOH, and add water to make the volume to 1 L. All the reagents are of ACS grade.
3. Immunoaffinity column (IAC). The column (AflaCLEAN™, LCTech GmbH) contains antibodies raised against aflatoxins B_1, B_2, G_1, and G_2. The column shall have a total capacity of 100 ng of aflatoxin B_1. The column should be stored at room temperature, in the dark (see Note 2).

2.2. Standard Stock and Working Solutions

1. Aflatoxins stock solution (500 μg/L). Prepare a 500 μg/L stock solution of standard aflatoxins (containing 200 μg/L of AFB_1, 50 μg/L of AFB_2, 200 μg/L of AFG_1, and 50 μg/L of AFG_2) from a standard solution of aflatoxins B_1, B_2, G_1, and G_2, with a total certified concentration of 5 μg/mL (corresponding to a concentration of 2.0 μg/mL each for B_1 and G_1, and of 0.5 μg/mL each for B_2 and G_2), supplied by Riedel-de Haën (Sigma-Aldrich, Laborchemikalien, Seelze,

Germany) (see Note 3). Pipette 1 mL of the 5 μg/mL certified solution into a 10-mL volumetric flask. Dilute to the mark with acetonitrile. This solution is stable for at least 6 months when stored at −20°C (see Note 4).

2. Aflatoxins solution (5 μg/L). Prepare an intermediate 5 μg/L aflatoxins stock solution (containing 2.0 μg/L of AFB_1, 0.5 μg/L of AFB_2, 2.0 μg/L of AFG_1, and 0.5 μg/L of AFG_2) by pipetting 100 μL of the 500 μg/L aflatoxins stock solution into a 10-mL volumetric flask. Dilute to the mark with mobile phase (water/acetonitile/methanol, 55/15/30). This solution is stable for 2 months when stored at −20°C, protected against light.
3. Aflatoxins standard solutions for HPLC. From the 5 μg/L aflatoxins solution, prepare four standard solutions into 2-mL amber vials, according to Table 1. Dilute each standard solution with mobile phase (water/acetonitrile/methanol 55/15/30, by volume).

2.3. Experimental Conditions of Chromatographic Separation and Fluorescence Detection

1. HPLC mobile phase (water/acetonitrile/methanol 55/15/30, by volume). Prepare a solution of water/acetonitrile/methanol (55/15/30, by volume) in a 1-L HPLC reservoir, by mixing 55 volume parts of water (550 mL) with 15 volume parts of acetonitrile (150 mL) and 30 volume parts of methanol (300 mL) (see Note 1).

2.4. Equipment

1. HPLC system. The chromatographic system consists of the following:
   (a) Binary pump equipped with a microvacuum degasser.
   (b) Thermostated autosampler.
   (c) Column compartment.
   (d) Fluorescence detector, fitted with a flow cell and set at 365 nm (excitation) and 435 nm (emission).

Table1
Preparation of aflatoxins standard solutions for HPLC

Standard solutions for HPLC	Volume (μL) from a 5 μg/L solution	Volume (μL) of mobile phase added	Final aflatoxin concentration (μg/L)			
			AFB_1	AFB_2	AFG_1	AFG_2
1	125	875	0.25	0.0625	0.25	0.0625
2	375	1,125	0.5	0.125	0.5	0.125
3	750	750	1.0	0.25	1.0	0.25
4	1,000	–	2.0	0.5	2.0	0.5

The system is interfaced, via a network chromatographic software, to a personal computer for the control of instruments, data acquisition, and processing.

2. Analytical reversed-phase separating column. ZORBAX Eclipse® XDB-C18 (150 mm×4.6 mm i.d., particle size 5 μm) from Agilent Technologies, kept at 40°C. Flow rate: 1.0 mL/min. Injection volume: 20 μL.
3. On-line photochemical post-column derivatization system. Commercially available UVE™ system supplied by LCTech GmbH (Dorfen, Germany) consisting of the following:
   (a) 254-nm low-pressure mercury lamp for the photochemical derivatization of aflatoxins.
   (b) 1-mL knitted reaction coil, fitted around the UV lamp.

3. Methods

3.1. Sample Extraction

1. Weigh 15.00 ± 0.01 g of grounded sample into a 50-mL polypropylene tube and add 30 mL of extraction solvent (MeOH/H_2O 80/20, v/v).
2. Vortex for a few minutes and then filter through paper, avoiding transferring of solid material on the filter.
3. Pipette 2 mL of the filtered extract into a flask, add 8 mL of PBS, and mix thoroughly. This solution will be cleaned by using the immunoaffinity column.

3.2. Immunoaffinity Column Clean-Up

1. Remove the top cap from the column and discard the storing buffer.
2. Pipette 10 mL of the filtered extract diluted in PBS in the column. Let the extract flow through the column at approximately 1–2 drops per second and discard the eluate. A slow, steady pressure is essential for the binding of the aflatoxins by the antibody.
3. Wash the column with 10 mL of water at a rate lower than 0.5 mL/min and discard the eluate.
4. Elute aflatoxins with 2 mL of methanol at a flow rate of no more than one drop per second or by gravity. Push air through the column to collect the entire eluate.
5. Evaporate the eluate to dryness under a stream of nitrogen at 40°C.
6. Dissolve the dried residue with 2 mL of mobile phase (water/acetonitrile/methanol 55/15/30, by volume) and filter through Anotop 10 LC (0.2 μm, Whatman) inorganic membrane.

3.3. HPLC-FLD Determination

The method to be used should be checked whenever the analysis of a different matrix is performed. For this purpose, comparing the chromatograms of blank (see Note 5) and spiked samples is essential in order to verify the absence of interfering peaks in the retention time-window of interest (±2.5% of the retention time of each aflatoxin).

3.4. Detection and Quantification of Aflatoxins

1. Identify aflatoxins B_1, B_2, G_1, and G_2 by comparing the retention times of each real sample with those of an aflatoxins standard solution. For example, in Fig. 1, typical chromatograms of an aflatoxins standard solution and a spiked wheat bran sample are shown.
2. Calibration curves for each aflatoxin can be obtained by three series of analyses on three different days, injecting standard solutions of aflatoxins B_1 and G_1, each at concentrations of 0.25, 0.5, 1.0, and 2.0 μg/L, and aflatoxins B_2 and G_2, each at concentrations of 0.0625, 0.125, 0.25, and 0.5 μg/L, prepared as described in Table 1. A good linearity with a correlation coefficient higher than 0.9995 is expected.
3. Detection (LOD) and quantification (LOQ) limits are calculated according to the equations $LOD = 3.3 \times s_a / b$ and

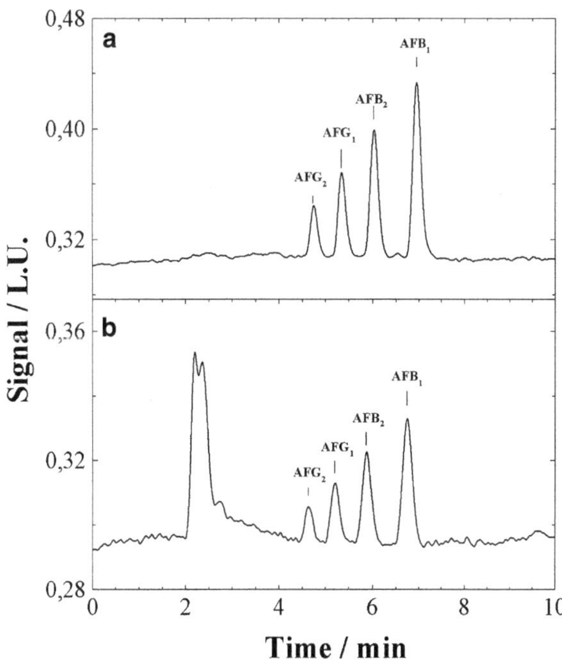

Fig. 1. Typical chromatograms of an aflatoxins standard solution at a total concentration of 5 μg/L (**a**) and a wheat bran sample spiked with AFB_1 and AFG_1 at 2.0 μg/kg and AFB_2 and AFG_2 at 0.5 μg/kg (**b**).

Table 2
Limits of detection and quantification of the method

	LOD		LOQ	
Aflatoxin	**μg/L**	**μg/kg**	**μg/L**	**μg/kg**
B_1	0.04	0.08	0.11	0.22
B_2	0.01	0.02	0.04	0.08
G_1	0.08	0.16	0.24	0.48
G_2	0.02	0.04	0.05	0.10

LOQ = $10 \times s_a/b$, where s_a is the standard deviation of the intercept and b is the slope of the regression line obtained from the calibration curves. The standard deviations of slope and intercept are estimated at the 95% confidence level. The expected limits of detection and quantification, evaluated by injecting aflatoxins standard solutions, and the corresponding matrix values are reported in Table 2. Calibration parameters should be checked at the beginning of each working session.

4. Quantify the concentrations of aflatoxins (reported in μg/L) in the eluate by comparing sample peak areas with the calibration curve. Calculate the corresponding concentration in the sample (μg/kg) by multiplying by the dilution factor 2, evaluated for each aflatoxin taking into account the sample extraction and clean-up.

3.5. Recovery and Precision

1. The method should be periodically tested for the evaluation of trueness and intra- and inter-day precision to ensure the agreement of the experimental data with provisions of Regulation (EC) No. 401/2006, dealing with the official control analyses of mycotoxins in foodstuffs (9).
2. Precision and recovery data are obtained from the analysis of samples fortified, prior to the extraction, with proper amounts of a standard solution of aflatoxins at final concentrations of 0.5, 1, and 1.5 times the maximum limit (ML) established in Regulation (EC) No. 1881/2006 (10).
3. The analyses should be performed on different days with the same instruments, but with different operators and instrumental calibrations. Recoveries of aflatoxins are evaluated by comparing the concentration of spiked samples determined by interpolation on the calibration curve with the nominal fortification level.

Table 3
Within-laboratory reproducibility and recovery evaluated for spiked wheat bran samples

Aflatoxin	Contamination Level (μg/kg)	Recovery % (mean value ± SD)	RSD_R % Intra-day (*n* = 6)	RSD_R % Inter-day (*n* = 18)
B_1	1	76.4 ± 6.5	3.5	9.9
	2	83.0 ± 9.8	4.9	11
	3	92.2 ± 8.3	2.7	6.8
B_2	0.25	79.5 ± 11.6	9.1	12
	0.50	82.1 ± 9.9	5.9	11
	0.75	92.8 ± 8.0	3.1	7.6
G_1	1	80.4 ± 9.6	7.7	14
	2	83.4 ± 11.1	6.2	15
	3	94.6 ± 8.2	4.4	8.1
G_2	0.25	89.2 ± 16.9	8.8	20
	0.50	82.0 ± 10.7	11	19
	0.75	87.7 ± 10.1	6.6	10

4. Typical precision and recovery data evaluated for each aflatoxin in spiked wheat bran samples at three fortification levels are summarized in Table 3. The values are derived from the validation procedure performed according to the European regulations (9, 11).

4. Notes

1. Use only reagents of recognized analytical grade in the analysis. All the solutions should be prepared in water that has a resistivity of 18.2 MΩ cm. Water and organic solvents should be of HPLC grade.
2. Do not freeze. The antibody of the immunoaffinity column can be denaturated by extreme conditions of temperature and pH. Excessive pressure on the column or extreme vacuum in the vacuum unit may cause compression of the gel and consequently low recovery.
3. Aflatoxins are very hazardous substances and have been classified as group 1 of human carcinogens by the International

Agency for Research on Cancer (IARC). Observe appropriate safety precautions for sample and standard handling. Suitable protective clothing, including rubber gloves, safety glasses, and laboratory coats, should be worn throughout the analysis. All materials, glassware, and toxin-containing solutions should be decontaminated before reusing or discarding by soaking in a 10% solution of sodium hypochlorite for about 18 h.

4. Aflatoxins are light sensitive; therefore, avoid exposure of the standard solutions and sample extracts to direct light. Original samples should be kept refrigerated, dry, protected against light, and well sealed.
5. Blank samples can be tested for the absence of aflatoxins by using common ELISA screening methods.

Acknowledgments

The authors gratefully acknowledge P. D'Antini and G. Battafarano (Istituto Zooprofilattico Sperimentale della Puglia e della Basilicata) for their technical assistance. This work was supported by Ministero della Salute (Rome, Italy).

References

1. Bullerman LB (1979) Significance of mycotoxins to food safety and human health. J Food Prot 42:65–86
2. Dickens F, Jones HE (1963) The carcinogenic action of aflatoxin after its subcutaneous injection in the rat. Br J Cancer 17:691–698
3. Garner RC, Kingfisher CN (1979) Chemical carcinogens and DNA, vol 1. CRC Press, Boca Raton
4. Trucksess MW, Stack ME (1994) Enzyme-linked immunosorbent assay of total aflatoxins B1, B2, and G1 in corn: follow-up collaborative study. J AOAC Int 77:655–658
5. Garcia-Villanova RJ, Cordón C, Gonzáles Paramás AM, Aparicio P, Garcia Rosales ME (2004) Simultaneous immunoaffinity column cleanup and HPLC analysis of aflatoxins and ochratoxin A in Spanish bee pollen. J Agric Food Chem 52:7235–7239
6. Tavčar-Kalcher G, Vrtač K, Pestevšek U, Vengušt A (2007) Validation of the procedure for the determination of aflatoxin B1 in animal liver using immunoaffinity columns and liquid chromatography with postcolumn derivatization and fluorescence detection. Food Control 18:333–337
7. Kussak A, Andersson B, Andersson K (1995) Determination of aflatoxins in hot chilli products by matrix solid-phase dispersion and liquid chromatography. J Chromatogr A 708:55–60
8. Muscarella M, Iammarino M, Nardiello D, Lo Magro S, Palermo C, Centonze D, Palermo D (2009) Validation of a confirmatory analytical method for the determination of aflatoxins B_1, B_2, G_1 and G_2 in foods and feed materials by HPLC with on-line photochemical derivatization and fluorescence detection. Food Addit Contam 26:1402–1410
9. European Commission (2006) Regulation (EC) No 401/2006. Off J Eur Union L 70:12–34
10. European Commission (2006) Regulation (EC) No 1881/2006. Off J Eur Union L 364:5–24
11. European Commission (2004) Regulation (EC) No 882/2004. Off J Eur Union L 165:1–141

Chapter 19

Highly Sensitive PCR-Based Detection Specific to *Aspergillus flavus*

Amaia González-Salgado, Teresa González-Jaén, Covadonga Vázquez, and Belén Patiño

Abstract

Aspergillus flavus is an important fungal species that frequently contaminates food commodities with diverse toxins, with aflatoxins being the most relevant in food safety. In addition, this is one of the major pathogenic *Aspergillus* species. In this work, specific PCR-based protocol for this species is described which allows the discrimination of other closely related species from the *Aspergillus* section *Flavi*, particularly *Aspergillus parasiticus*. The specific primers were designed on the multicopy internal transcribed region of the rDNA unit (ITS1-5.8S-ITS2 rDNA).

Key words: *Aspergillus flavus*, Aflatoxin, Polymerase chain reaction, Detection, Internal transcribed spacer

1. Introduction

Aflatoxins are potent carcinogenic, mutagenic, and teratogenic secondary metabolites and are produced predominantly by *Aspergillus flavus* and *Aspergillus parasiticus* (1, 2). Currently, upper limits of aflatoxins in foodstuffs are under regulation in the European Union (3). This regulation often results in the destruction of contaminated agricultural products, causing significant economic losses. Aflatoxigenic fungi can contaminate several food commodities including cereals (4), peanuts (5), spices (6), and figs (7, 8). In addition, *A. flavus*, together with *Aspergillus fumigatus*, are responsible for 90% of aspergillosis in human beings (9).

The level of mold infestation and the identification of governing species are important indicators of the quality of the raw material, and predict the potential risk for the presence of mycotoxins (10).

Otto Holst (ed.), *Microbial Toxins: Methods and Protocols*, Methods in Molecular Biology, vol. 739, DOI 10.1007/978-1-61779-102-4_19,

Early detection of aflatoxin-producing species is crucial to prevent aflatoxins entering the food chain. PCR-based methods that target DNA are considered a good choice for rapid diagnosis because of their high specificity and sensitivity, especially when multicopy target sequences are used (11). This PCR protocol is based on multicopy sequences (ITS of rDNA units) specific to *A. flavus*, and allows a rapid, very efficient, and sensitive detection of this fungus.

2. Materials

2.1. DNA Extraction

1. DNeasy Plant Mini Kit (Qiagen). Store at room temperature (15–25°C), except RNase A and AW buffer (at 4°C), and AP1 (at 28°C to avoid precipitation of components).
2. Ethanol (96–100%). Store at room temperature.

2.2. PCRs

1. Primers FLA1 (5′-GTAGGGTTCCTAGCGAGCC-3′) and FLA2 (5′-GGAAAAAGATTGATTTGCGTTC-3′) (Sigma-Aldrich): prepare aliquots of each primer at 20 μM in sterile water and store at –20°C.
2. PCR buffer (10×), $MgCl_2$ solution (50 mM), and Taq DNA polymerase (5 U/mL) (Biotools). Store at –20°C.
3. dNTPs solution (100 mM) (Biotools). Store at –20°C.
4. Molecular biology grade water (MoBio, Carlsbad, USA).
5. Template DNA (10–200 ng). Store at –20°C.

2.3. Agarose Gel Electrophoresis

1. TAE buffer (1×): Tris–acetate 40 mM and EDTA 1.0 mM. Store at room temperature.
2. Agarose D-1 Low EEO (Pronadisa, Madrid, Spain). Store at room temperature.
3. 1% Ethidium bromide solution (Applichem). Store at room temperature.
4. Loading buffer: 25 mg bromophenol blue, 4 g sucrose, and H_2O to 10 mL. Store at 4°C.
5. 100-bp DNA molecular mass marker (MBI Fermentas, Vilnius, Lithuania). Store at 4°C.

3. Methods

3.1. DNA Extraction

1. These instructions assume the extraction of DNA from pure fungal culture (see Note 1), contaminated plant material, or food matrix (see Note 2). The DNeasy Plant Mini Kit includes RNase A enzyme, buffers, columns, and collection tubes.

2. Prepare in advance all the materials required for the protocol: sterilize all material (tips, mortar and pestle, microcentrifuge tubes, etc.); add the appropriate amount of ethanol to buffer AW and AP3/E, as indicated on the bottle, to prepare a working solution; preheat a water bath to 65°C; and label each microcentrifuge tube and column.
3. Grind the plant, food matrix, or fungal tissue under liquid nitrogen using a sterile mortar and pestle (see Note 3). Transfer the tissue powder (a maximum of 100 mg) to a 1.5-mL microcentrifuge tube.
4. Add 400 μL of buffer AP1 (lysis buffer) and 4 μL of RNase A stock solution (100 mg/mL) and vortex vigorously.
5. Incubate the mixture for 10 min at 65°C to lyse the cells. Mix two or three times during incubation by inverting the tube.
6. Add 130 μL of buffer AP2 (precipitation buffer, see Note 4) to the lysate, mix, and incubate for 5 min in ice to precipitate detergent, proteins, and polysaccharides present in the sample.
7. Centrifuge for 6.5 min at 16,000 × *g*.
8. Transfer the supernatant into the QIAshredder Mini spin column (lilac) placed in a 2-mL collection tube, and centrifuge for 2.5 min at 16,000 × *g*.
9. Transfer the flow-through fraction (usually about 430 μL) from step 8 into a new tube without disturbing the cell debris pellet.
10. Add 1.5 volume (usually about 645 μL) of buffer AP3/E (binding buffer) to the cleared lysate, and mix immediately by pipetting.
11. Pipette 650 μL of the mixture from step 10, including any precipitate that may have formed, into the DNeasy Mini spin column placed in the 2-mL collection tube. Centrifuge for 1 min at 6,000 × *g* and discard the flow-through. Reuse the collection tube in step 12.
12. Repeat step 11 with remaining sample. Discard the flow-through and collection tube.
13. Place the DNeasy Mini spin column into a new 2-mL collection tube, add 500 μL of buffer AW (wash buffer), and centrifuge for 1 min at 6,000 × *g*. Discard the flow-through and reuse the collection tube in step 14.
14. Add 500 μL of buffer AW to the DNeasy Mini spin column and centrifuge for 2.5 min at 16,000 × *g* to dry the membrane.
15. Transfer the DNeasy Mini spin column to a 1.5-mL or 2-mL microcentrifuge tube and add 100 μL of molecular degree water directly onto the DNeasy membrane. Incubate for

5 min at room temperature and then centrifuge for 1 min at 6,000 × g to elute.

16. Measure the amount of template DNA spectrophotometrically. The concentration and purity of DNA can be determined by measuring the absorbance at 260 nm (A260) and 280 nm (A280). Purity is determined by calculating the ratio of absorbance at 260 nm to absorbance at 280 nm. Pure DNA has an A260/A280 ratio of 1.8–2.0.
17. Store the DNA at −20°C.

3.2. PCR Amplification

1. Mix the following components in a 0.5-mL microcentrifuge tube: 2 μL (10–200 ng total DNA amount, see Note 5) of template DNA (see Note 6), 1 μL of each primer FLA1 and FLA2, 2.5 μL of 10× PCR buffer, 1 μL of $MgCl_2$, 0.2 μL of dNTPs, and 0.15 μL of Taq DNA polymerase (see Note 7). Add molecular biology grade water up to 25 μL.
2. It is recommended to perform control PCR assays (see Note 8).
3. The PCR amplification protocol is 1 cycle of 5 min at 95°C; 26 cycles of 30 s at 95°C (denaturalization), 30 s at 58°C (annealing), and 45 s at 72°C (extension); and finally, 1 cycle of 5 min at 72°C.
4. Store the PCR products at 4°C.

3.3. Agarose Gel Electrophoresis

1. These instructions are for making a 10 × 15 cm 1% agarose gel using a Bio-Rad Wide Mini-Sub Cell GT gel system.
2. Weigh out 0.6 g of agarose into a 250-mL conical flask. Add 60 mL of 1× TAE and swirl to mix.
3. Microwave the mixture for about 1 min to dissolve the agarose.
4. Leave it to cool down to about 60°C (just hot to hold in bare hands).
5. Add 1 μL of ethidium bromide (10 mg/mL) and swirl to mix (see Note 9).
6. Pour the gel slowly into the tray. Insert the comb and check that it is correctly positioned. The gel should polymerize within 20 min.
7. Pour 1× TAE buffer into the gel tank to submerge the gel to 2–5 mm depth. This is the running buffer.
8. Prepare the samples by adding 5 μL of loading buffer to a 15 of each sample into an appropriate labeled tube.
9. Once the gel has set, remove the comb carefully, and load 20 μL of each sample in a well. Reserve one well for the DNA molecular mass marker.

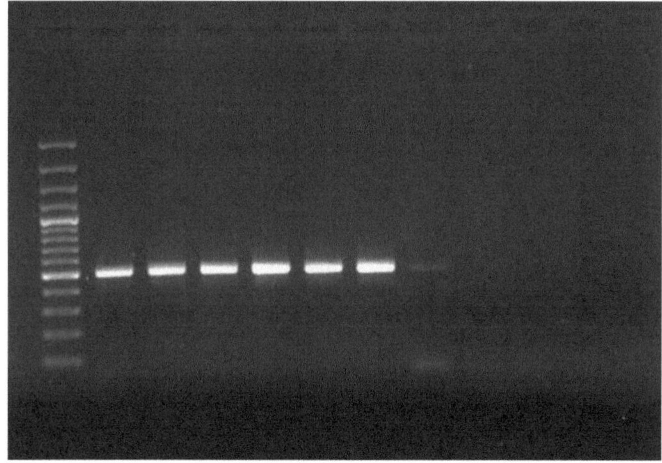

Fig. 1. PCR amplification using primers FLA1/FLA2 and DNA from Lanes 1 and 2: *Aspergillus flavus* strains (CECT 2681, ITEM 4591); Lanes 3 and 4: DNA from wheat flour contaminated with *A. flavus*; Lane 5: DNA from barley flour contaminated with *A. flavus*; Lanes 6 and 7: DNA from paprika contaminated with *A. flavus*; Lane 8: Non-template control; Lane 9: *Aspergillus parasiticus* Cab5dch6; Lane 10: *Aspergillus bombycis* IHEM 19289; Lane 11: *Aspergillus carbonarius* CECT 2086. M: 100-bp DNA molecular mass marker.

10. Complete the assembly of the gel unit and connect to a power supply. Run the gel at 80 V for about 45 min while monitoring the progress of the gel with reference to the marker dye.
11. The gel is then placed in an UV-light box to visualize the PCR products. An example of the results produced is shown in Fig. 1.

4. Notes

1. Fungal DNA can be obtained from isolates cultured in 100-mL Erlenmeyer flasks containing 20 mL of liquid Sabouraud medium (Scharlau Chemie, Barcelona, Spain) and incubated at 28°C and 150 rpm. *A. flavus* mycelia from 3-day-old cultures can be harvested by filtration through Whatman No. 1 paper and frozen in liquid nitrogen. Mycelia can be stored at –80°C until DNA extraction.
2. This protocol can be adapted for many other systems, using instead other more appropriate DNA extraction kits, e.g., DNeasy Blood & Tissue kit (Qiagen).
3. The primary disruption of the tissue is an essential step that must be performed carefully. It can be performed with another disruption method, such as commercially available homogenizers.

4. Phenolic compounds present in some food matrices, such as grapes or pepper, might interfere with DNA extraction procedure or inhibit PCR assay. In these cases, polyvinylpyrrolidone (0.33%) (Sigma-Aldrich) can be added to AP2 buffer in order to remove these compounds.
5. When the assay is performed from a fungal pure culture, the total amount of DNA obtained is normally between 10 and 50 ng. When PCR assay is performed with DNA from a food matrix or a plant tissue, the total amount of DNA per reaction should be between 150 and 200 ng.
6. Positive (with 2 μL of *A. flavus* pure DNA) and negative (with 2 μL of molecular biology grade water instead of DNA template) controls must be included in PCR assays.
7. Use filter tips to prepare PCR. Taq polymerase must be maintained in ice during its manipulation.
8. It is recommended to test the DNA integrity by performing a control PCR with the universal fungal primers ITS1 (5′ TCCGTAGGTGAACCTGCGG 3′) and ITS4 (5′ TCCTCCGCTTATTGATATG 3′), using the amplification program described by Henry et al. (12).
9. Ethidium bromide is a potent mutagenic agent and must be handled with care; always wear a suitable laboratory coat and disposable gloves. All ethidium bromide waste must be disposed in appropriate containers.

References

1. Bennett JW, Klich M (2003) Mycotoxins. Clin Microbiol Rev 16:479–516
2. Horn BW (2007) Biodiversity of *Aspergillus* section *Flavi* in the United States: a review. Food Addit Contam 24:1088–1101
3. Commission Regulation (2006) EC No 1881/2006 setting maximum levels for certain contaminants in foodstuffs. Official Journal of the European Union, Brussels
4. Pittet A (1988) Natural occurrence of mycotoxins in foods and feeds-an update review. Rev Med Vet-Toulouse 149:479–492
5. Jelinek CF, Pohland AE, Wood GE (1989) Worldwide occurrence of mycotoxins in foods and feeds-an update. J Assoc Off Anal Chem 72:223–230
6. Bartine H, Tantaoui-Elaraki A (1997) Growth and toxigenesis of *Aspergillus flavus* isolates on selected spices. J Environ Pathol Toxicol Oncol 16:61–65
7. Doster MA, Michailides TJ, Morgan DP (1996) *Aspergillus* species and mycotoxins in figs from California orchards. Plant Dis 80:484–489
8. Färber P, Geisen R, Holzapfel WH (1997) Detection of aflatoxinogenic fungi in figs by a PCR reaction. Int J Food Microbiol 36:215–220
9. Paya CV (1993) Fungal infection in solid organ transplantation. Clin Infect Dis 16:677–688
10. Shapira R, Paster N, Eyal O, Menasherov M, Mett A, Salomon R (1996) Detection of aflatoxinogenic molds in grains by PCR. Appl Environ Microbiol 62:3270–3273
11. González-Salgado A, González-Jaén MT, Vázquez C, Patiño B (2008) Highly sensitive PCR-based detection method specific for *Aspergillus flavus* in wheat flour. Food Addit Contam 25:758–764
12. Henry T, Iwen P, Hinrichs S (2000) Identification of *Aspergillus* species using internal transcribed spacer regions 1 and 2. J Clin Microbiol 38:1510–1515

Chapter 20

A Rapid Enzymatic Method for Aflatoxin B Detection

Danila Moscone, Fabiana Arduini, and Aziz Amine

Abstract

A novel method for aflatoxin B (AFB) determination is proposed. The AFB determination is based on acetylcholinesterase (AChE, from *electric eel*) inhibition, and the AChE residual activity is determined using the colorimetric method (Ellman's method). To select and optimize the analytical procedures, the investigation on type of AChE inhibition by AFB_1 was carried out. The AChE degree of inhibition by AFB_1 was independent of the incubation time and the enzyme concentrations, showing the reversibility of the inhibition. This reversibility of the inhibition permits a rapid analysis of AFB_1. In fact, only a 3-min analysis is required. For the development of AFB_1 assay, the pH, the reaction time, the temperature, and the substrate concentration were evaluated and optimized. The linear range of 10–60 ng/mL was assessed. To evaluate the selectivity of this method, the cross-reactivity with other aflatoxins, such as AFB_2 (aflatoxin B_2), AFG_1 (aflatoxin G_1), AFG_2 (aflatoxin G_2), and AFM_1 (aflatoxin M_1), was investigated. The suitability of the assay for AFB_1 quantification in barley was also evaluated. This study shows a new approach to detect aflatoxins based on enzyme inhibition with several advantages, such as the easiness of use, the rapidity, and the cost-effectiveness, demonstrating a possible use as screening method for this type of mycotoxins.

Key words: Cholinesterase, Aflatoxin, Inhibition, Ellman's method, Bioassay

1. Introduction

The family of aflatoxins has been shown to represent a significant class of mycotoxins owing to their documented impact on both human and animal health (1, 2). Due to its ability of covalent binding to DNA and proteins (3–5), aflatoxin B_1 (AFB_1) is the most acutely and chronically toxic member of the aflatoxin family.

The current reference analytical methods are primarily chromatographic, such as high-performance liquid chromatography (HPLC) (6–8), that, however, require specialized laboratory, skilled personnel, high cost, and are not suitable for "in situ"

Otto Holst (ed.), *Microbial Toxins: Methods and Protocols,* Methods in Molecular Biology, vol 739, DOI 10.1007/978-1-61779-102-4_20, © Springer Science+Business Media, LLC 2011

application. In contrast, enzymatic methods have usually shown promise as an alternative to classical methods to achieve a faster and simpler detection of some environmental pollutants, such as pesticides, heavy metals, etc. (9–15). In the case of the aflatoxins, this approach can be explored to find an enzyme that can be inhibited by AFB_1. In a recent study, Cometa et al. (16) have analyzed the inhibition of acetylcholinesterase (AChE) by AFB_1, a key enzyme in nervous impulse. The AFB_1 inhibition on AChE extracted from *mouse brain* was studied and the implications in terms of kinetic mechanism and toxicity discussed. In this study, we have developed an analytical method that allows AFB_1 determination based on AChE (from *electric eel*, commercially available) inhibition by this toxin. On the basis of this effect, a method to quantify AFB_1 by measuring the decrease in enzyme activity was developed. The AChE activity was evaluated using Ellman's spectrophotometric method with a detection of AFB_1 in barley (17).

2. Materials

2.1. Aflatoxin Solutions

Aflatoxins B_1, B_2, G_1, G_2, and M (Vinci-Biochem, Italia) were dissolved in methanol and stored in aliquots at –20°C. Working solutions are prepared by dilution with methanol. Stock solutions of aflatoxins (highly toxic compounds) were prepared under appropriate safety conditions, i.e., in order to avoid contact with the powder or inhalation of vapor of aflatoxins, the operators were protected with lab dresses, gloves, mask, and glasses. Also, a fume hood has to be used during sample preparation and analysis.

2.2. Enzyme Solution

The AChE from *electric eel* (EC 3.1.1.7) was prepared in water and stored in aliquots of 40 U/mL at –20°C.

2.3. Enzymatic Substrate Solution

Acetylthiocholine chloride (ATCh, Sigma) solutions were in general prepared by dissolving in water. The solutions must be prepared freshly every day and, during the day, maintained at 4°C.

2.4. Ellman's Method

1. Spectrophotometer Unicam 8625 UV/VIS.
2. Cuvette made of PVC (1.5 mL).
3. Phosphate buffer 0.1 M, pH 8, stored at 4°C and adjusted to room temperature (25°C) before use.
4. DTNB [5,5′-dithiobis-(2-nitrobenzoic acid)] (Sigma Chemical Company, St. Louis, MO) is dissolved in phosphate buffer 0.01 M, pH 7, by stirring for several hours until the solution appears to be a yellow homogeneous solution. It is stored at 4°C and adjusted to room temperature before use.

2.5. Study of pH Effect

1. Phosphate buffers, 0.1 M, pH 6, 7, and 8, stored at 4°C and adjusted to room temperature before use.
2. Tris–HCl buffer, 0.1 M, pH 9, stored at 4°C and adjusted to room temperature before use.

2.6. AFB$_1$ Extraction

1. Shaker (e.g., M 101-OR, Instrument, Italy).
2. Rotavapor (e.g., V.V. MICRON, Heidolph, Germany).
3. Centrifuge (e.g., ALC MOD. PK 120).

3. Methods

The analytical method for AFB$_1$ detection developed in this research work is based on AChE inhibition by this toxin. AChE, a key enzyme in nervous transmission, catalyzes the hydrolysis of the neurotransmitter acetylcholine to choline and acetic acid. In the presence of compounds that are able to inhibit this enzyme, the enzyme activity decreases and the amount of the decrease can be correlated to the inhibitor concentration. In order to monitor the enzymatic activity, the Ellman's method was used. In this case, a nonnatural substrate was used (acetylthiocholine) that allows the formation of the enzymatic products thiocholine and acetic acid. The thiocholine is then able to react with the Ellman's reagent (DTNB (5,5′-dithiobis-2-nitrobenzoic acid)] with the consequent production of TNB (2-nitro-5-thiobenzoic acid), a yellow compound with a maximum absorbance peak at 412 nm (18) (Fig. 1).

This method allows the detection of any inhibitor of AChE in a very easy, fast, and economic way. Briefly, in the absence of an inhibitor a yellow color is observed and the absorbance is measured spectrophotometrically, whereas in the presence of the inhibitor, a colorless solution is obtained.

3.1. Measurement of Cholinesterase Activity and Michaelis–Menten Constant

The AChE activity is determined by measuring the product of the enzymatic reaction. The ACTh was chosen as substrate and the thiocholine content, enzymatically produced by AChE, was identified applying Ellman's spectrophotometric method.

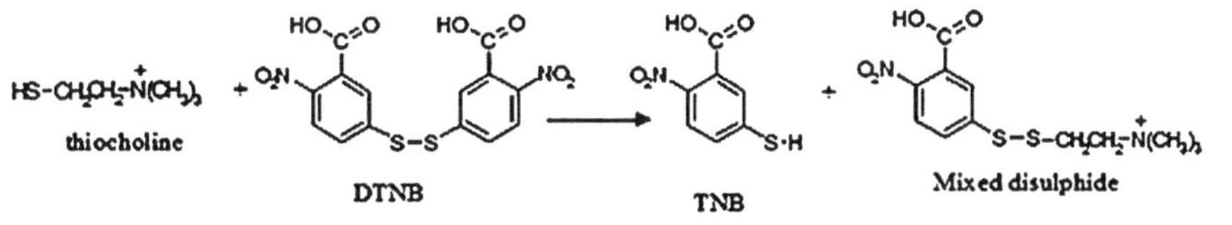

Fig. 1. Scheme of the Ellman's reaction.

Enzymatic activity measurement

Step a

1. 900 μL of phosphate buffer (0.1 M, pH 8) are added to a cuvette.
2. 100 μL of 0.1 M DTNB are then added to the same cuvette.
3. The cuvette is inserted in a spectrophotometer to perform a blank measurement; usually, the absorbance is less than 0.100. This value is subtracted automatically by the spectrophotometer from the values obtained in further measurements carried out during the entire working day.

Step b

4. To a new cuvette, 872 μL of phosphate buffer (0.1 M, pH 8) and 100 μL of 0.1 M DTNB are added.
5. 20 μL of AChE solution (2 U/mL) are added to the cuvette.
6. 8 μL of ATCh at the selected concentration are put into the cuvette. The solution is then stirred utilizing a Gilson pipette and the reaction time immediately starts, followed by a timer. After 150 s, the solution in the cuvette is stirred again and after 180 s the value of absorbance is read.

In order to determine the Michaelis–Menten constant (K_M), different amounts of ACTh are added to the cuvette to obtain final concentrations of 0.1, 0.2, 0.3, 0.4, 0.5, 0.6, 1, 2, and 5 mM.

The graph in Fig. 2 shows an evident behavior of enzymatic kinetic described by Eq. 1 in which V is the reaction rate, V_{max} is the maximum reaction rate, and (S) is the substrate concentration.

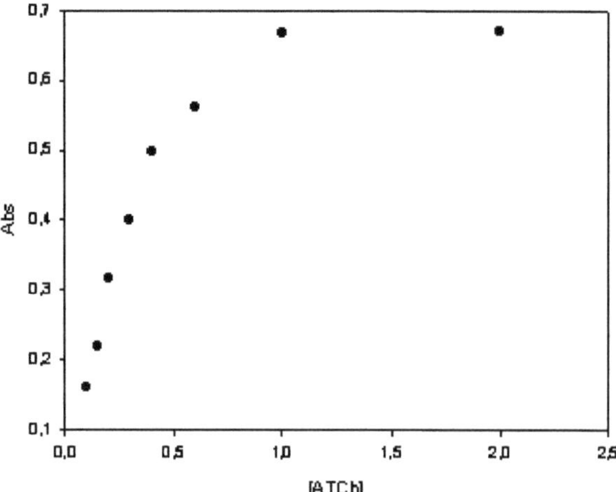

Fig. 2. Calibration plot of acetylthiocholine. AChE = 40 mU mL^{-1}, 3 min of reaction time, phosphate buffer 0.1 M, pH 8.

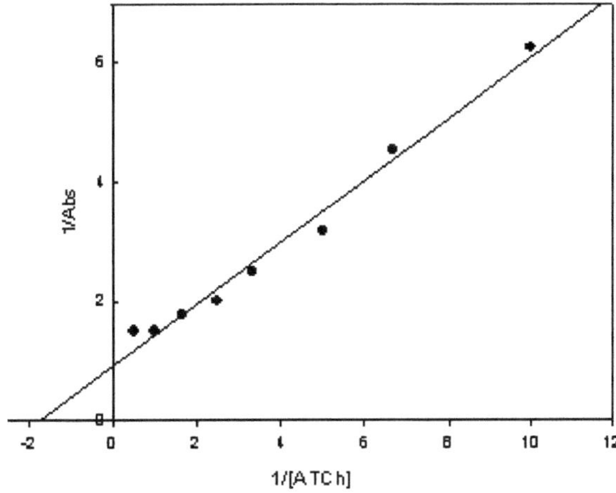

Fig. 3. Lineweaver-Burk plot representing reciprocals of the initial enzyme velocity vs. ATCh concentration. AChE = 40 mU mL^{-1}, 3 min of reaction time, phosphate buffer 0.1 M, pH 8.

$$V = (V_{\text{max}} \times S)/(K_{\text{M}} + S) \qquad (1)$$

K_{M} is the Michaelis–Menten constant which describes the affinity of the enzyme toward its substrate and corresponds numerically to the substrate concentration at 1/2 V_{max}.

Then, the data are plotted using a Lineweaver Burk's plot (19), correlating $1/V$ (the 1/values of absorbance) against the 1/concentration of the substrate in Fig. 3.

This graph is described by Eq. 2:

$$1/V = 1/V_{\text{max}} + (K_{\text{M}}/V_{\text{max}}) \times 1/S \qquad (2)$$

From the graph, it is possible to calculate V_{max} from the intercept, and K_{M} from the slope, equal to 0.26 mmol/L min and 0.33 mM, respectively.

3.2. Measurement of the Methanol Effect

3.2.1. Measurement of the Methanol Effect on Ellman's Method

Before starting the investigation of AChE inhibition by AFB_1, a study on the effect of the organic solvent used to solubilize the AFB_1 on the enzymatic activity is necessary. In fact, several organic solvents can affect any enzymatic activity which holds also true for cholinesterase enzymes (20, 21).

Methanol is often used to extract AFB_1 from various contaminated agricultural samples (22–24); thus, methanol was chosen as organic solvent to solubilize AFB_1 in our experiments.

First, the effect of methanol on Ellman's method is evaluated. Cysteamine is taken as a representative thiol for this investigation, since it is commercially available, whereas thiocholine has to be enzymatically produced (25).

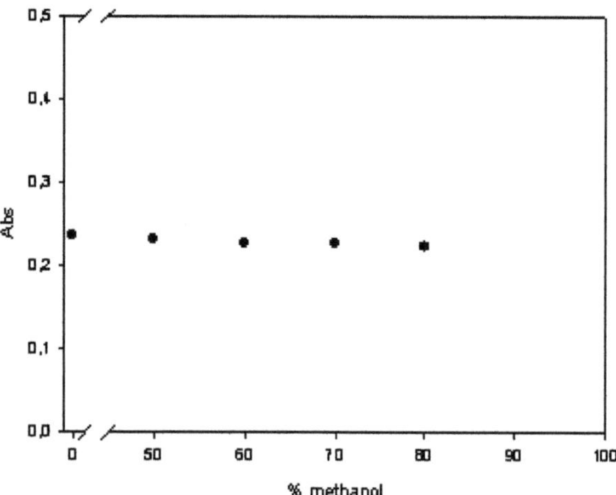

Fig. 4. Effect of methanol (percentage from 0% up to 80%) on Ellman's reaction, cysteamine 1.9 mM, phosphate buffer 0.1 M, pH 8.

Procedure 1

1. 900 μL of phosphate buffer (0.1 M, pH 8) are added to a cuvette.
2. 100 μL of 0.1 M DTNB are then added.
3. The cuvette is inserted in a spectrophotometer to perform a blank measurement; usually, the absorbance is less than 0.100. This value is subtracted automatically by the spectrophotometer from the values obtained in the further measurements carried out during the entire working day.
4. To a new cuvette, $(890 - M)$ μL of phosphate buffer (0.1 M, pH 8), M μL of methanol, and 100 μL of 0.1 M DTNB are added.
5. 10 μL of cysteamine (1.9 mM) are added to the cuvette in order to obtain a final concentration of 1.9×10^{-5} M. The solution is mixed and the absorbance measured (M represents the μL of methanol added to yield the different percentages illustrated in Fig. 4).

As shown in Fig. 4, there should be no effect of methanol on Ellman's method.

3.2.2. Measurement of the Methanol Effect on Enzyme Activity

It is important to test the influence of methanol on the enzymatic activity, because, as largely reported in literature (20, 21), the presence of organic solvents generally utilized for the extraction of analytes may affect the enzymatic activity leading to doubtful results.

Procedure 2

Step a

1a. 900 μL of phosphate buffer (0.1 M, pH 8) are added to a cuvette.

2a. 100 μL of 0.1 M DTNB are added.

3a. The cuvette is inserted in a spectrophotometer to perform a blank measurement, usually the absorbance is less than 0.100. This value is subtracted automatically by the spectrophotometer from the values obtained in the further measurements carried out during the entire working day.

Step b

1b. To a new cuvette, 872 μL of phosphate buffer (0.1 M, pH 8) are added.

2b. 20 μL of AChE (2 U/mL) are added.

3b. Then, 8 μL of 0.05 M ATCh are added, and the solution is stirred utilizing a Gilson pipette. The reaction time immediately starts and is followed by a timer. After 165 s, 100 μL of 0.1 M DTNB are added to the cuvette and the solution is again stirred with the Gilson pipette.

4b. After 180 s (total time), the absorbance is read and this value corresponds to the enzyme activity in the absence of methanol (A_0).

Step c

1c. To a new cuvette, $872 - M$ μL of phosphate buffer (0.1 M, pH 8) and M μL of methanol are added (M corresponds to the microliters of methanol added to the cuvette for obtaining different percentage of methanol, from 0 to 90% v/v).

2c. 20 μL of AChE (2 U/mL) are added.

3c. Then, 8 μL of 0.05 M ATCh are added to the solution which is then stirred utilizing a Gilson pipette. The reaction time immediately starts and is followed by a timer. After 165 s, 100 μL of 0.1 M DTNB are added in the cuvette and the solution is stirred with the Gilson pipette.

4c. After 180 s, the value of absorbance is read and this value corresponds to the enzyme activity in the presence of methanol (A_i).

The inactivation of the enzyme caused by the exposure to methanol is calculated by Eq. 3, in which $I\%$ is the percentage of enzyme inactivation due to methanol, A_0 is the absorbance obtained in aqueous solution, and A_i is the absorbance obtained in aqueous solution with methanol.

$$I\% = \left(A_0 - A_i\right)/A_0 \tag{3}$$

Figure 5 shows that the AChE activity decreases in parallel to the increase in the amount of methanol. At 50% methanol (in phosphate buffer), the AChE activity should have decreased by 30%.

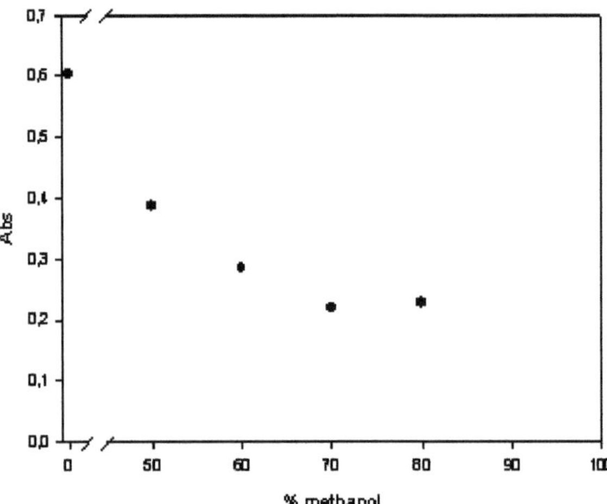

Fig. 5. Effect of methanol (percentage from 0% up to 80%) on AChE activity. AChE = 40 mU mL^{-1}, 3 min of reaction, acetylthiocholine = 0.4 mM, phosphate buffer 0.1 M, pH 8.

If the percentage of methanol is increased to 80%, the inactivation seemed to level off at 60% of the original activity. However, it was found that in the range 0–5% of methanol added, no relevant inactivation is observed. Thus, in setting up the inhibition assay system, less than 5% should always be used (see Note 1).

3.3. Investigation of the Inhibition Mechanism of AFB on AChE

Since little is known about the inhibition by AFB_1 (2, 16), it is relevant to understand the type of inhibition. This study allows to optimize all the analytical parameters in an easier and faster way, and to identify the lower detection limit.

The procedure could also be applied to study the type of inhibition of other enzymes inhibited by other compounds.

3.3.1. The Reversibility of the AChE Inhibition by AFB1

First, it is relevant to understand whether the inhibition is irreversible or reversible. In the first case, it is important to use a low enzyme concentration in order to obtain a low limit of detection while in the case of reversible inhibition, no change in the degree of inhibition should be observed using different enzyme concentrations (10, 26). The degree of inhibition at fixed concentration of AFB_1 (60 ng/mL), using various concentrations of AChE, was then determined. The enzyme concentrations used for this experiment were 70, 40, and 7 mU/mL. For this purpose, the following procedure was utilized:

Procedure 3 (See Note 2)

Step a

1a. 900 μL of phosphate buffer (0.1 M, pH 8) are added to a cuvette.

2a. 100 μL of 0.1 M DTNB are then added.

3a. The cuvette is inserted in a spectrophotometer to perform a blank measurement; usually, the absorbance is less than 0.100. This value is subtracted automatically by the spectrophotometer from the values obtained in the further measurements carried out during the entire working day.

Step b

1b. To a new cuvette, [(900−*X*−*Y*−*M*), where M are microL of methanol] μL of phosphate buffer (0.1 M, pH 8) are added.

2b. Then, *X* μL of AChE (2 U/mL) and 8 μL of 0.05 M ATCh are added.

3b. Immediately, the solution is stirred utilizing a Gilson pipette, and the reaction time starts and is followed by a timer. After 165 s, 100 μL of 0.1 M DTNB are added and the solution is again stirred with the Gilson pipette.

4b. After 180 s, the value of absorbance is read corresponding to the enzyme activity in the absence of AFB_1 (A_0).

Step c

1c. In a new cuvette, (900 − *X* − *Y* − *W*) μL of phosphate buffer (0.1 M, pH 8), and *W* μL of aflatoxin at *Z* concentration are added.

2c. *X* μL of AChE at (2 U/mL) concentration are added and the solution stirred.

3c. After *N* min, *Y* μL of ATCh are added and the solution is stirred utilizing a Gilson pipette, and the reaction time starts and is followed by a timer. After 165 s, 100 μL of 0.1 M DTNB are added and the solution is again stirred with the Gilson pipette.

4c. After 180 s, the value of absorbance is read which corresponds to the enzyme activity in absence of AFB_1 (A_i) in which

   X (μL of 2 U/mL AChE at added to the cuvette) is equal to 35, 20, and 3.5 in order to have a final concentration in a cuvette equal to 70, 40, and 7 mU/mL, respectively;

   Y (μL of 0.05 M acetylthiocholine added to the cuvette) = 8;

   M (μL of methanol in the cuvette) = 30;

   N (incubation time) = 0; *Z* (concentration of aflatoxin) = 2 mg/L

   W (μL of aflatoxin added) = 30.

Applying Eq. 3, in which A_0 are the values obtained by step b and A_i those of step c, for the enzyme concentrations 70, 40, and 7 mU/mL, the degrees of inhibition should be 45 ± 3%, 50 ± 4%, and 47 ± 3%, respectively.

These results (essentially no change in degree of inhibition) apparently support the hypothesis that the inhibition of AChE by AFB_1 follows a reversible mechanism. To confirm this hypothesis, a study of the incubation time with the inhibitor is performed. While for the case of irreversible inhibition, an increase in incubation time is required to reach a low detection limit, for reversible inhibition a longer incubation time does not lead to an increase in the degree of inhibition. Then, the effect of incubation time [the time of reaction between the inhibitor (AFB_1) and the enzyme (AChE)] is also evaluated using procedure 3 but with X=20 μL AChE (2 U/mL) and incubation times (N) of 0, 1, 2, 3, 5, 10, and 30 min.

Applying Eq. 3, in which A_0 are the values obtained from step b and A_i those from step c, the degree of inhibition is calculated. For the experiments performed at the different incubation times, the same degree of inhibition was obtained, i.e., 48 ± 3%. Given that the incubation time has no effect on the degree of inhibition, and the interaction is rapid and reversible, there is no need for preincubation of the enzyme with the inhibitor before the substrate is added to start the reaction.

3.3.2. Study of the Type of AChE Inhibition by AFB_1

The study of the inhibition mechanism is an important point that should be carried out before every bioassay development, except if this mechanism is published. By knowing this mechanism, it is possible to identify those parameters that have to be optimized in order to reach the lowest detection limit.

Similar experiments are carried out to evaluate whether the inhibition is competitive. In this case, the concentration of substrate affects the degree of inhibition and then, to reach lower detection limit, it is necessary to use the lowest concentration of substrate that allows a detectable signal in a reasonable time. AChE activity was determined using ATCh concentrations over the range from values near $K_M/4$ up to $4K_M$, either in the absence or in presence of fixed AFB_1 concentrations (60 ng/mL).

For this purpose, procedure 3 was used with X=20 μL AChE (2 U/mL), 26.4, 13.2, and 6.6 μL of 0.05 M ATCh (Y) for $4K_M$, $2K_M$, and K_M, respectively, and at an incubation time N=0. In the case of $K_M/4$ and $K_M/2$, the stock solution of ATCh was diluted ten times and from this, a volume of 16.5 and 33 μL was added to the cuvette. Applying Eq. 20.3 (A_0, values obtained from step b; A_i, values from step c), the degree of inhibition was calculated to 47 ± 5%, indicating that the inhibition is not competitive in nature. Generally, the amount of enzymatic products depends on both, the concentration of the enzyme and that of the substrate, and on the reaction time. Knowing that the inhibition is not competitive in nature, it is possible to use high concentrations of substrate such as equal to $4K_M$ without changing the result of inhibition, but decreasing the time of analysis.

To understand in detail which type of noncompetitive inhibition (uncompetitive, noncompetitive, or of the mixed type) is present in this case, the AChE activity is determined with the previous range of ATCh concentration (values near $K_M/4$ to

$4K_M$) either in the absence or in presence of different AFB_1 concentrations. These experiments are necessary to construct the Lineweaver–Burk plot with and without inhibitor, which is used to understand the type of inhibition mechanism.

The experiments are performed using the procedure 3 with the following parameters:

1. Measurement of enzymatic activity in absence of AFB_1: X=20 μL; N=0; Z=0; M=0; W=0; Y=26.4, 13.2, and 6.6 μL for the measurement of enzymatic activity in the case of substrate concentration equal to $4K_M$, $2K_M$ and K_M. In the case of an enzymatic activity with a substrate concentration equal of $K_M/4$ and $K_M/2$, the stock solution of 0.05 M ATCh is diluted ten times (10 μL of 0.05 M ATCh in 90 μL of water) and from this, 16.5 and 33 μL are added in the cuvette, respectively.
2. Measurement of the enzymatic activity in the presence of 75 μg/L of AFB_1: the same parameters as in the point 1 but with W=37.5 μL, M=37.5 μL, and Z=2 mg/L.
3. Measurement of enzymatic activity in the presence of 300 ng/mL AFB_1: the same parameters were used as in the point 1 but with W= 15 μL, M= 15 μL, and Z=20 mg/L.

The Lineweaver–Burk plot is constructed following the procedure described in Subheading 3.1 in the absence and presence of 75 and 300 ng/mL of AFB_1 (Fig. 6). The lines obtained pass through different points on the ordinate and intersect at a point slightly displaced from the abscissa axis in the second quadrant.

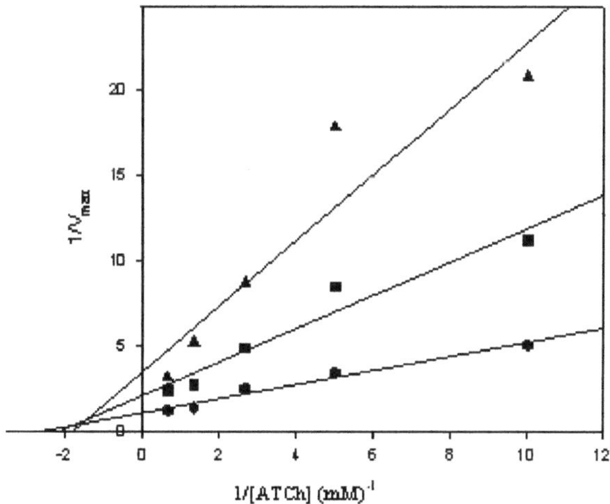

Fig. 6. Lineweaver-Burk plot representing reciprocals of the initial enzyme velocity vs. ATCh concentration without (●) and with various concentrations of AFB_1 (75 ■ and 300 ▲ ng mL^{-1} of AFB_1). AChE = 40 mU mL^{-1}, 3 min of reaction, phosphate buffer 0.1 M, pH 8.

3.4. Optimization of the AFB_1 Assay Based on AChE Inhibition

In order to develop an effective assay for AFB_1 determination based on the decrease in enzyme activity, Ellman's spectrophotometric method for the determination of AChE activity has to be adapted and optimized, taking into consideration parameters such as substrate concentration, solvent, pH, etc.

As noted above, the characteristics of the inhibition make it possible to optimize the enzyme and substrate concentration of the assay mixture in order to achieve suitable assay times and levels of inhibition. An enzyme concentration of 40 mU/mL is chosen because it allows the achievement of reasonable optical densities from Ellman's reaction within only 3 min.

Given that the incubation time has no effect on the degree of inhibition, there is no need for a prior incubation of the enzyme with the inhibitor before adding the substrate to initiate the reaction. For this reason, no incubation time is required for the rest of the work. Thus, the optimized parameters of the procedure 3 are $X=20$ μL, $N=0$, and $Y=8$ μL.

3.4.1. Effect of pH

As reported in our previous work (27), the best buffer and pH value for AChE is 0.1 M phosphate buffer at pH 8. To evaluate the AChE inhibition by AFB_1 as a function of pH, AChE, ATCh, and AFB_1 are added to buffers with different pH applying procedure 3 with $X=20$ μL, $N=0$, $Y=8$ μL, $W=30$ μL, $M=30$ μL and $Z=2$ mg/L, using for each experiment the buffer $[(900-X-Y-W)$ μL$]$ at selected pH.

The degrees of inhibition obtained are shown in Fig. 7. We observed the highest degree of inhibition with phosphate

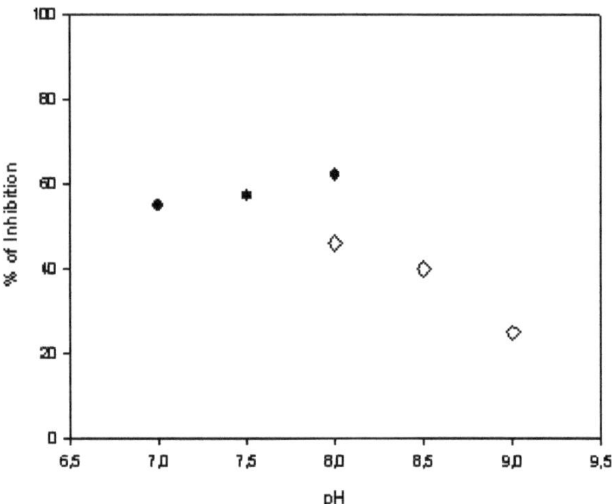

Fig. 7. Effect of pH on AChE inhibition by 60 ng mL^{-1} of AFB_1. AChE = 40 mU mL^{-1}, 3 min of reaction, ATCh = 0.4 mM. At pH 7, 7.5 and 8 the phophate buffer 0,1 M (◆) was used while at pH 8, 8.5 and 9 a Tris buffer 0,1 M (◇) was chosen.

buffer at pH 8, and a rapid decrease in Tris–HCl buffer at pH 9 is observed. Thus, phosphate buffer at pH 8 is chosen for further experiments.

3.4.2. Effect of Temperature on Enzyme Activity in Storage Condition

Storage stability, or shelf life, is referred to an enzyme maintaining catalytic properties in the period between manufacture and eventual use. Temperature is an important parameter that influences the enzyme stability (28). The effect of the temperature on the enzyme activity and on the inhibition of AChE by AFB_1 was evaluated using procedure 3 and the optimized parameters (Subheading 3.4) with $Z = 2$ mg/L and $W = 30$ μL.

As can be seen in Fig. 8a, AChE stored at room temperature is gradually inactivated while no decrease of enzymatic activity was observed when the enzyme was maintained at 4°C. On the contrary, the effect of temperature on the inhibition is more evident (Fig. 8b); thus, the thawed enzyme can be used only for 1 working day and stored, for the entire day, at 4°C.

3.4.3. Effect of Methanol on Enzyme Inhibition by AFB_1

The effect of methanol on the inhibition of AChE by AFB_1 is also investigated using the procedure 3 (Subheading 3.4) with $Z = 2$ mg/L varying the percentage of methanol from 0 to 50%. A degree of inhibition of 60 ng/mL of AFB_1 ($50 \pm 3\%$) should be obtained. This result demonstrates that even the highest percentage of methanol has no effect on the degree of AChE inhibition by AFB_1, although such amount of methanol decreases the level of AChE activity. Thus, it is possible to determine AFB_1 using a percentage of methanol as high as 50%, allowing to dilute the AFB_1 standard solution (prepared in pure methanol) twofold.

3.4.4. Calibration Curve

Five successive calibration curves were performed in the range between 10 and 60 ng/mL and the results give a linear correlation: $y = (13.4 \pm 0.9) + (0.602 \pm 0.009)x$, where y is the degree of inhibition and x is the AFB_1 concentration with r^2 equal to 0.955. The lower limit of the linear range, defined as the concentration giving an inhibition of 20%, was 10 ng/mL of AFB_1 while the IC_{50} was 60 ng/mL of AFB_1.

To establish the calibration curve, procedure 3 was used, adding also AFB_1 in step c and methanol in step b. The calibration curve was constructed using six different AFB_1 concentrations (10, 20, 30 ,40, 50, and 60 ng/mL). The parameters are listed in Table 1.

3.4.5. Cross-Reactivity

To evaluate the selectivity of the proposed assay, IC_{50} values are determined using the other aflatoxins (AFB_2, AFG_1, AFG_2, and AFM_1) for comparison with AFB_1. Procedure 3 is adopted for this purpose. The investigation is carried out applying the parameters reported in Table 1. In this case, several concentrations (Z) and

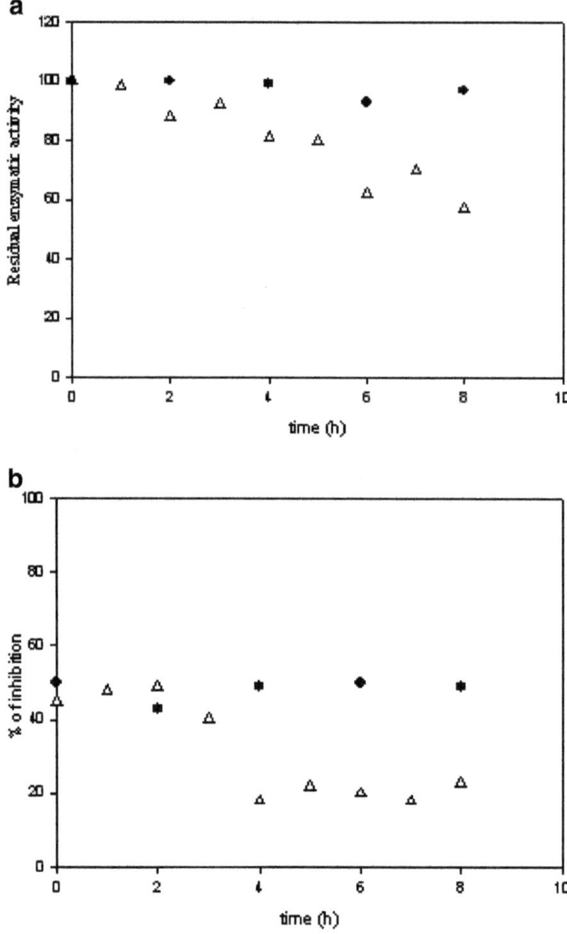

Fig. 8. Effect of temperature on AChE storage conditions. AChE = 40 mU mL^{-1}, 3 min of reaction time, ATCh = 0.4 mM (a) and with also AFB_1 60 ng mL^{-1} (b). The residual activity (fig 8a) and the degree of inhibition (fig 8b) were evaluated for AChE maintained at room temperature (△) and at 4°C (✱).

volumes (*W*) of the different aflatoxins tested are added to the cuvette in order to obtain a degree of inhibition around 50%.

As shown in Table 2, both AFB_1 and AFB_2 provoke considerable inhibition of AChE while both AFG_1 and AFG_2 show limited capacity to inhibit the enzyme. The results for AFM_1 are intermediate with an IC_{50} of 175 ng/mL. The results obtained show that the assay can detect AFB_1 and AFB_2 with almost the same sensitivity.

3.5. AFB_1 Measurement in Fortified Barley Samples

The AChE inhibition assay is then applied for the determination of AFB_1 using spiked barley samples in order to test its performance in a real matrix. In the first step, the matrix effect is evaluated using barley without AFB_1. The sample is treated following procedure 4 (29):

Table 1
Parameters used to construct the calibration curve

Parameters	Step b	Step c
X (μL of AChE at 2 U/mL added to the cuvette)	20	20
N (incubation time)	0	0
Z (concentration of aflatoxin)	0	2 mg/L
W (μL of aflatoxin added to the cuvette)	0	5 for 10 ng/mL 10 for 20 ng/mL 15 for 30 ng/mL 20 for 40 ng/mL 25 for 50 ng/mL 30 for 60 ng/mL
Υ (μL of 0.05 M acetylthiocholine added to the cuvette)	8	8
M (μL of methanol added to the cuvette)	5 for 10 ng/mL 10 for 20 ng/mL 15 for 30 ng/mL 20 for 40 ng/mL 25 for 50 ng/mL 30 for 60 ng/mL	0

Table 2
Concentration of aflatoxin resulting in 50% enzyme inhibition (IC_{50})

Aflatoxin	IC_{50} (ng/mL)	*W* (μL)	*Z* (mg/mL)
AFB_1	60	30	2
AFB_2	72	36	2
AFG_1	850	17	50
AFG_2	310	15.5	20
AFM_1	175	26.5	6.6

W = μL of aflatoxin at *Z* concentration added to the cuvette, AChE = 40 mU/mL, 3 min of reaction time, and ATCh = 0.4 mM

1. 5 g of barley are milled.
2. The barley powder obtained is mixed for 15 min with 25 mL of a solution of methanol (70%), phosphate buffer (29%), and *N*,*N*-dimethylformamide (1%).
3. The mixture is centrifuged at 2,000 × *g* for 15 min.
4. The supernatant is dried by a rotavapor at 30°C (around 20 min).

5. The AFB_1 is then solubilised in 1 mL of methanol.
6. The solution is then filtered using a glass filter.

   After that, the solution obtained is diluted (1:15 v/v) in a cuvette, and calibration curves (Fig. 9) are obtained applying the procedure in Subheading 3.4.4. Thus, the only experimental change is in the initial solution which is $833 - X - Y + 67$ μL of barley solution. In this way, the matrix effect was evaluated.

The extraction efficiency of the toxin from barley is calculated using samples spiked with known amounts of the toxin prior to the extraction. In this way, a 200-μL aliquot of AFB_1 standard solution at different concentrations is added to 5 g of blank powder barley in order to obtain 100, 120, and 150 ng/g of AFB_1 in barley sample. After that, the treatment of the sample is performed as described in procedure 4. On the basis of the calibration curves prepared with barley extract (Fig. 9, filled circle), it is possible to calculate the extraction efficiency (Table 3) of the analyte. The results show that

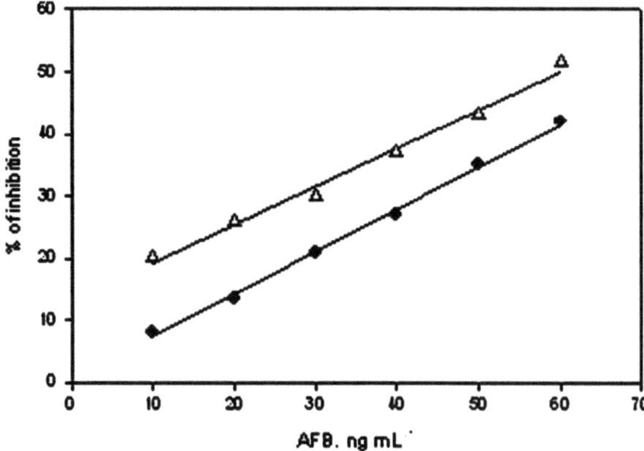

Fig. 9. Calibration curves performed in phosphate buffer 0.1 M (△) and in barley matrix diluted 1/15 (v/v) (◆). AChE = 40 mU mL^{-1}, 3 min of reaction, ATCh = 0.4 mM.

Table 3
Recovery studies of spiked barley samples using assay for AFB_1 based on AChE inhibition (concentration in barley)

AFB_1 added (ng/mL)	AFB_1 found (ng/mL)	% Recovery
150	152	101
120	117	98
100	99	99

The data reported in the table represent the average of three measurements (reproduced from (17))

the procedure adopted for extraction of AFB_1 from barley is quick and shows recovery of 99 ± 5%.

4. Conclusion

A new method for AFB_1 detection was developed. The determination of AFB_1 is based on its capacity to inhibit AChE from *electric eel*, utilizing Ellman's spectrophotometric method to measure the enzyme activity in the absence and presence of analyte. In fact, within a few minutes, it is possible to obtain results suitable for rapid screening of the prepared samples.

If it can be demonstrated that the inhibition of AChE by aflatoxins bears a direct relation to their toxic effects (e.g., cholinergic ones), then the method has the additional advantage of effectively detecting the presence of harmful substances in order to remove contaminated lots from the food production chains.

A preliminary investigation of the kinetics and type of inhibitory mechanism indicated a mixed type. It is interesting to note that the binding of AFB_1 to the *electric eel* enzyme is about three orders of magnitude stronger than that found for the mouse brain AChE (16), and this behavior allowed AFB_1 to be detected, by the enzyme inhibition-based assay, at ng/mL levels. The LOD achieved was 10 ng/mL, rendering this method useful for the screening of AFB_1 in food. Finally, the cross-reactivity results showed that the assay detected AFB_2 with similar sensitivity to that for AFB_1 while this method had low capacity to detect AFM_1, AFG_1, and AFG_2.

5. Notes

1. The percentage of methanol added to a cuvette is a key step in this method. As demonstrated in the text, an increase of the percentage of methanol causes a decrease of the enzymatic activity, due to the enzyme inactivation by methanol, but we have also demonstrated that the degree of inhibition is independent of the percentage of methanol up to 5%. In case of the calibration curve, different concentrations of AFB_1 must be added to the cuvette. In this procedure, it is important to add *always* a known volume of AFB_1 (dissolved in methanol) in step c and the same volume of methanol in step b, in order to have *always* the same percentage of methanol in both steps.

2. In case of aflatoxin determination, procedure 3 was adopted taking into consideration that the DTNB (Ellman's reagent) should be added only 15 s before the absorbance is read. In fact, we have experimentally observed that if the DTNB is added at the beginning of the measurement, as reported in procedure 1, the inhibition is lower and not reproducible. In our hypothesis, this behavior is due to the interaction between AFB_1 and DTNB through π–π interactions, however, this hypothesis needs to be confirmed by further investigations.

References

1. Bennett JW, Klich M (2003) Mycotoxins. Clin Microbiol Rev 16:497–516
2. Hussein HS, Brasel JM (2001) Toxicity, metabolism and impact of mycotoxins on humans and animals. Toxicology 167:101–134
3. Iwaki M, Kitagawa T, Akamatsu Y, Aibara K (1990) Cytotoxic effects of aflatoxin B1 and its association with cellular components in chicken embryo primary cultured cells. Biochim Biophys Acta 1035:146–153
4. Cole KE, Jones TW, Lipsky MM, Trump BF, Hsu IC (1988) In vitro binding of aflatoxin B_1 and 2-acetylaminofluorene to rat, mouse and hepatocyte DNA: the relationship of DNA binding to carcinogenity. Carcinogenesis 9:711–716
5. Wilson DM, Payne GA (1994) The toxicology of aflatoxin, human health veterinary and agricultural significance. Academic, San Diego
6. Blesa J, Soriano JM, Moltò JC, Marin R, Manes J (2003) Determination of aflatoxin in peanuts by matrix solid-phase dispersion and liquid chromatography. J Chromatogr 1011:49–54
7. Pearson SM, Candlish AAG, Aidoo KE, Smith JE (2004) Determination of aflatoxin levels in pistachio and cashew nuts using immunoaffinity column clean-up with HPLC and fluorescence detection. Biotechnol Tech 13:97–99
8. Elizalde-Gonzalez MP, Mattusch J, Wennich R (1998) Stability and determination of aflatoxins by high-performance liquid chromatography with amperometric detection. J Chromatogr 828:439–444
9. Yang Y, Yang M, Wang H, Tang L, Shen G, Yu R (2004) Inhibition biosensor for determination of nicotine. Anal Chim Acta 509:151–157
10. Amine A, Mohammadi H, Bourais I, Palleschi G (2006) Enzyme inhibition-based biosensors for food safety and environmental monitoring. Biosens Bioelectron 21:1405–1423
11. Sezginturk MK, Goktug T, Dinckaya E (2005) A biosensor based on catalase for determination of highly toxic chemical azide in fruit juices. Biosens Bioelectron 21:684–688
12. Ghica ME, Brett CMA (2008) Glucose oxidase inhibition in poly(neutral red) mediated enzyme biosensors for heavy metal determination. Microchim Acta 163:185–193
13. Shan D, Mouty C, Cosnier S (2004) Subnanomolar cyanide detection at polyphenol oxidasse/clay biosensors. Anal Chem 76:178–183
14. Campas M, Marty JL (2007) Enzyme sensor for the electrochemical detection of the marine toxin okadaic acid. Anal Chim Acta 605:87–93
15. Cosnier S, Mousty C, Cui X, Yang X, Dong S (2006) Specific determination of As(V) by an acid phosphatase polyphenol oxidase biosensor. Anal Chem 78:4985–4989
16. Cometa MF, Lorenzini P, Fortuna S, Volpe MT, Meneguz A, Palmery A (2005) *In vitro* inhibitory effect of aflatoxin B1 on acetylcholinesterase activity in mouse brain. Toxicolgy 206:125–135
17. Arduini F, Errico I, Amine A, Micheli L, Palleschi G, Moscone D (2007) Enzymatic spectrophotometric method for aflatoxin B detection based on acetylcholinesterase inhibition. Anal Chem 79:3409–3415
18. Ellman GL (1959) Tissue sulfhydryl groups. Arch Biochem Biophys 82:70–77
19. Lineawever H, Burk D (1934) The determination of enzyme dissociation constants. J Am Chem Soc 56:658–666
20. Mionetto N, Marty JL, Karube I (1994) Acetylcholinesterase in organic solvents for the detection of pesticides: biosensor application. Biosens Bioelectron 9:463–470
21. Iwuoha EI, Smyth MR, Lyons MEG (1997) Organic phase enzyme electrodes: kinetics and

analytical applications. Biosens Bioelectron 12:53–75

22. Bourais I, Amine A, Venanzi M, Micheli L, Moscone D, Palleschi G (2005) Development and application of a two-phase clean-up system in food samples prior to fluorescence analysis of aflatoxins. Microchim Acta 539:195–201
23. Lin L, Zhang J, Wang P, Wang Y, Chen J (1998) Thin-layer chromatography of mycotoxins and comparison with other chromatographic methods. J Chromatogr A 815:3–20
24. AOAC Official Method (1994) 991.31 A
25. Arduini F, Ricci F, Bourais I, Amine A, Moscone D, Palleschi G (2005) Extraction and detection of pesticides by cholinesterase inhibition in a two-phase system: a strategy to avoid heavy metal interference. Anal Lett 38:1703–1719
26. Arduini F, Amine A, Moscone D, Palleschi G (2009) Reversible enzyme inhibition based biosensors: applications and analytical improvement through diagnostic inhibition. Anal Lett 42:1258–1293
27. Arduini F, Ricci F, Tuta C, Moscone D, Amine A, Palleschi G (2006) Detection of carbamic and organophosphorous pesticides in water samples using cholinesterase biosensor based on Prussian Blue modified screen printed electrode. Anal Chim Acta 580:155–162
28. Peterson ME, Daniel RM, Danson MJ, Eisenthal R (2007) The dependence of enzyme activity on temperature: determination and validation of parameters. Biochem J 402:331–337
29. Ammida NHS, Micheli L, Palleschi G (2004) Electrochemical immunosensor for determination of aflatoxin B1 in barley. Anal Chim Acta 520:159–164

INDEX

Otto Holst (ed.), *Microbial Toxins: Methods and Protocols*, Methods in Molecular Biology, vol. 739,
DOI 10.1007/978-1-61779-102-4, © Springer Science+Business Media, LLC 2011

T

V

Z

MIX
Papier aus verantwortungsvollen Quellen
Paper from responsible sources
FSC® C105338

If you have any concerns about our products,
you can contact us on
ProductSafety@springernature.com

In case Publisher is established outside the EU,
the EU authorized representative is:
Springer Nature Customer Service Center GmbH
Europaplatz 3, 69115 Heidelberg, Germany

Printed by Libri Plureos GmbH
in Hamburg, Germany